MAMA LEARNED US TO WORK

STUDIES IN RURAL CULTURE | JACK TEMPLE KIRBY, EDITOR

LU ANN JONES

Mama Learned Us to Work

FARM WOMEN IN THE NEW SOUTH

The University of North Carolina Press | Chapel Hill & London

Designed by April Leidig-Higgins
Set in Monotype Garamond by Copperline Book Services, Inc.
Manufactured in the United States of America

The paper in this book meets the guidelines for permanence and
durability of the Committee on Production Guidelines for Book
Longevity of the Council on Library Resources.

Page ii: Lurline Stokes Murray standing in her chicken yard.
(Photo by Eric Long, 87-16469-9, Office of Imaging, Printing, and
Photographic Services, Smithsonian Institution, Washington, D.C.)

Portions of this book have been published in modified form elsewhere. Chapter 1 first appeared in "Gender, Race, and Itinerant
Commerce in the Rural New South," *Journal of Southern History* 66
(May 2000): 297–320. Part of Chapter 5 first appeared in "In
Search of Jennie Booth Moton, Field Agent, AAA," *Agricultural History* 72 (Spring 1998): 446–58. Chapter 6 is a version of an article
coauthored with Sunae Park, "From Feed Bags to Fashion," *Textile
History* (Spring 1993): 91–103. They are reprinted here with per-
mission of the publishers.

Library of Congress Cataloging-in-Publication Data
Jones, Lu Ann. Mama learned us to work: farm women in the
New South / Lu Ann Jones.
p. cm.—(Studies in rural culture)
Includes bibliographical references and index.
ISBN 0-8078-2716-9 (cloth: alk. paper)
ISBN 0-8078-5384-4 (pbk.: alk. paper)
1. Women in agriculture—Southern States. 2. Rural women—
Southern States. I. Title. II. Series.
HD6077.2.U6 J66 2002   305.43'63—dc21   2001059761

cloth   06 05 04 03 02   5 4 3 2 1
paper   06 05 04 03 02   5 4 3 2 1

In memory of my parents,
and for Bill

CONTENTS

ILLUSTRATIONS

Every woman has a farm woman in her
family and most of us do not have to go
back very far to find that woman.
—Joan M. Jensen, "Recovering Her Story"

Preface

MY MOTHER WAS the farm woman in my family, and her story began in
1920 on a farm in northeastern North Carolina. As a child, she studied by a
kerosene lamp, rode to school on a bus that routinely mired in the mud,
picked cotton, hoed the generous garden that her mother grew, and listened
to big band music on a battery-operated radio. She excelled in English, and
when she graduated from high school in 1937, my mother's ambition was to
attend business school and become a secretary. She rented a typewriter and
completed some of the courses at home. Just as she was preparing to go to
Norfolk, Virginia, to finish classes, her mother suffered a stroke. My mother,
the youngest daughter of the family, turned nurse and household manager.
By the time my grandmother recuperated, Mama had lost interest in typing
and shorthand. In June of 1941 she married my father, moved to his home-
place eight miles away, and began her life as a farmer's wife. Together my parents shaped and responded to the changes that defined southern rural life
during the next forty years.

Little did I know as a child in the 1950s and 1960s that the stories my mother told me would set a research agenda in years to come. Her stories were peopled with hucksters and the "Rawleigh man" and the "Watkins man." She described sewing with colorful feedbags. She belonged to the Corapeake Home Demonstration Club.

Like much of the work that I have done on farm women and southern agriculture in the past two decades, writing this book has kept me in dialogue with my own past. In 1983, my father died and my mother rented out the farm and moved to the town where my older brother lives. For the first time in five generations, no Joneses farmed our land. The next year I left the South altogether. From my new home in Connecticut, I watched as the agricultural crisis of the 1980s wreaked havoc in the countryside. With each farm foreclosure, I grew more convinced of the need to document the transformations in southern agriculture since the 1930s. The women and men who had farmed the land and endured those changes were growing older, and the time remaining to hear their stories was growing shorter. Fortunately, I found an intellectual home at the Smithsonian Institution, which supported my project, "An Oral History of Southern Agriculture," a dream come true. Conducting interviews with members of my parents' generation helped me to negotiate grief and endure loss during a time of disjuncture.

In the meantime, my mother continued to be one of my best sources. She inspired avenues of inquiry and suggested interpretations of information I was gathering. More comfortable writing me a letter than talking into my tape recorder, she amplified anecdotes and answered more detailed questions that I put to her. Her letter-essays about raising chickens, "the Watkins man," and feed bags occupy an important place in my files.

In the summer of 1991, just before I returned to North Carolina to complete my doctorate, Mama died. She was gone, but her stories remained. In part, this book is a search for their meanings. On so many counts, my mother opened my eyes to aspects of rural life that a scholar tutored by books alone might have overlooked or dismissed. Her stories pointed the way for me to follow in her footsteps while wearing my own shoes. My mother let me leave the farm without abandoning it.

I HAVE HAD THE GOOD fortune of receiving support from many people and institutions as I completed this work. I am grateful to the two hundred farm women and men throughout the South who shared their life stories with me between 1986 and 1991. Pete Daniel, curator of the defunct Division of Agriculture and Natural Resources at the National Museum of American History, made the research possible, and every day at work was a pleasure. At the museum, I learned from a number of gifted fellows and staff members. Jane Becker, Nancy Bercaw, Marjorie Berry, Jennifer Brier, Nanci Edwards, Francis Gadsden, Janet Greene, Christine Hoepfner, Louis Hutchins, Charlie McGovern, Grace Pallidino, Mary Panzer, Sunae Park, Nina Silber, Kim Wallace, Jeannie Whayne, and Nan Woodruff shared ideas and good times.

When I returned to graduate school at the University of North Carolina at Chapel Hill, a new cohort of students adopted me as one of their own. Pamela Grundy, Tom Hanchett, Laura Moore, and Marla Miller shared their knowledge of southern history and women's history. Members of my dissertation committee gave me the freedom to figure out the stories I wanted to tell and how I wanted to tell them. As a mentor, Jacquelyn Hall is unparalleled; her generosity as a scholar is legendary. Her passion and commitment taught me that history matters.

As I turned to revisions, several people took time to read and comment on all or portions of the manuscript. I owe special thanks to Grey Osterud, who challenged me to think more expansively about my work and who guided me over the rough spots that all writers encounter. Jack Temple Kirby's comments helped me rethink key sections. Anastatia Sims asked pointed questions and polished sentences as she read multiple drafts. I also wish to thank Peggy Barlett, Laura Edwards, Louis Kyriakoudis, Susan Levine, Mary Murphy, Ted Ownby, Adrienne Petty, and Mark Schultz for their suggestions.

At the University of North Carolina Press, Lewis Bateman expressed an early interest in my work. Chuck Grench was enthusiastic about a project he inherited and patiently awaited its completion. Suzanne Bell edited the manuscript with care, and Pam Upton shepherded it through production.

Archivists at many repositories have guided me to sources. I am especially grateful to George Stevenson at the North Carolina Division of Archives and History and to Maurice Toler and Caroline Weaver at the North Carolina State University Archives.

Friends and family have helped in countless ways over the years. For inspiring me to think capaciously about being a historian, I am grateful to Rob Amberg, David Cecelski, Harlan Gradin, Kathy Nasstrom, Kathy Newfont, Alicia Rouveral, and Maury York. My brothers, Allan Jones and Jimmy Jones, and my cousin, Page Styles, never gave up on me. Finally, my husband, Bill Mansfield, restored my soul day after day. Bill listened to every word that I wrote as we developed a ritual of readings. No writer could ask for more than the attention he gave me. Even as he took my work seriously, Bill never let me stop laughing. His love sustains me.

Writers and researchers need their patrons, and I have enjoyed generous financial support at every stage of this project. The Smithsonian Institution's Scholarly Studies Fund, the Women's Committee, and the James C. Smithson Society underwrote my project, "An Oral History of Southern Agriculture." While writing my dissertation, I received a Smithsonian Institution Graduate Student Fellowship, a UNC Graduate School On-Campus Dissertation Fellowship, Mowry Graduate Research Awards from the UNC Department of History, and an Albert J. Beveridge Research Grant from the American Historical Association. At East Carolina University, I was able to finish research and begin revisions with support from the Herbert R. Paschal Memorial Fund in the Department of History, a Research and Creative Activity Grant from the Division of Research and Graduate Studies, and a College Research Award from the College of Arts and Sciences. A fellowship from the National Endowment for the Humanities allowed me to complete the book.

MAMA LEARNED US TO WORK

Introduction

A CURIOUS THING happens when American women's history textbooks reach the twentieth century: farm women disappear. Stories of women on the land—Native American agriculturalists, colonial goodwives, plantation mistresses and slave mothers, New England farmers' daughters – turned – mill girls, and sojourners on westward trails—animate much of the narrative from the sixteenth through the nineteenth centuries. Then rural women drop out of sight, only returning for a cameo appearance during the 1930s when their care-worn faces evoke the suffering of the Great Depression. After the turn of the twentieth century, farm women are banished to the margins of women's history.[1]

Although the South remained the most rural region in the nation, historians of southern women have done little better. Just as scholars were once guilty of using New England women to represent American women, so the history of southern women in the twentieth century has suffered from its focus on those who broke with the past and moved to factories and towns. Only recently have regional studies of rural southern women joined sociologist Margaret Jarman Hagood's documentary classic, *Mothers of the South: Portraiture of the White Tenant Farm Woman*, published more than sixty years ago. The countryside has served primarily as backdrop and staging ground for black and white women who abandoned fields to work in tobacco factories or textile mills or to swell a new town-dwelling middle class. Overlooked were the thousands of women who stayed behind and continued to labor on farms. Neglected, too, was the part that farm and rural women of both races played in regional reform efforts that blossomed in the early twentieth century.[2]

By leaving out farm women, we suggest that little changed for women on the land, that they were outside of history. We perpetuate rather than interrogate the notion of a static, even "backward" and "traditional," rural society and culture. Because historians are usually preoccupied with the most dramatic ruptures with the past, we have failed to appreciate the ways in which rural women could be negotiating complex changes even as much about their lives remained the same during the first half of the twentieth century. Slighting farm women also dispossesses those of us who were born in the country of a vital part of our heritage.[3]

This book looks at the history of the rural South from the perspective of white and black farm women who have remained hidden in plain sight. A primary goal has been to move southern farm women from the sidelines to the center of history and to help them tell part of their story. That story is far more complicated than an image of a beleaguered woman to which they have all too often been reduced in the best of textbooks. And it challenges the stereotype of southern farm women as worn out by work and childbearing, worn down by masculine authority and racial oppression, and isolated by regional and rural geography.

MUCH OF THE INSPIRATION for this book came from interviews that I conducted with older rural southerners for "An Oral History of Southern Agriculture." Most of the two hundred narrators were born in the first two decades of the twentieth century and came of age during the Great Depression, as farming and rural life perched on the cusp of enormous changes.

Members of a watershed generation, they had grown up in a world of labor-intensive agriculture and grown old in the world of agribusiness. Project field-work followed the region's major commodity cultures. I talked to peanut farmers in southeastern Virginia and northeastern North Carolina. Farmers who raised cotton and flue-cured tobacco mingled in the coastal plain and piedmont of the Carolinas and Georgia, and cotton growers dominated in Mississippi's delta. Arkansas and Louisiana farmers shared their knowledge of rice and sugar. In the hill country of East Tennessee, north Georgia, and western North Carolina, farmers discussed an economy dominated by live-stock and grains.[4]

When I began conducting interviews, I carried with me a story of south-ern agriculture as presented in the scholarly literature that went something like this: Early twentieth-century farmers inherited a system that had been born after the Civil War and turned former slaves into sharecroppers. Credit based on crop liens tethered farmers to cash-crop production, pulled them into a market economy over which they had little control, and reduced an in-creasing number of white farmers from yeomen to tenants and sharecrop-pers. Most farmers followed the seasonal demands of cotton and tied their fates to the verdict reached at gins in the fall. A smaller group worked to-bacco, whose cultivation was so demanding from the time seeds sprouted until the last bundle of the golden leaf was sold that farmers joked that it took thirteen months a year to grow the crop. On the outskirts of the region, farmers on Arkansas's Grand Prairie raised rice and those in southwestern Louisiana grew sugarcane. Only mountain farmers maintained diversified farms oriented toward subsistence. Otherwise, the household economy had withered or died altogether after farm families entered the orbit of cash-crop production.

During the first two decades of the twentieth century, commodity prices remained generally favorable. While about half of the region's farmers worked someone else's land, the percentage of renters and sharecroppers leveled off. Then, in the 1920s, it was as if farmers had been standing on a trapdoor that suddenly opened. The bottom fell out of crop prices, and an economic de-pression settled over the countryside. More than ever, poverty became the hallmark of the rural South. In the 1930s, New Deal farm programs, with their acreage allotments and agricultural subsidy payments, began to fracture this system and set in motion another wave of change. Many poor black and white farmers were forced off the land when landlords refused to share the government's largesse and began to invest in tractors and implements. Others hung on during this reconfiguration of agriculture as machines and chemicals replaced human labor, the commodity mix changed, the federal government played a growing role in farm decision making, and survival came to rest on cash flow.[5]

As I saw it, my task as an interviewer was to help people describe these broad economic and social changes in personal terms and to interpret the interplay between structural changes and family and community life. Because I wanted to share interpretive authority with narrators, I was ready for them to challenge and complicate the received scholarly wisdom. I also began the fieldwork determined to listen to women as well as to men. In the 1980s, scholars were starting to take gender into account as they examined transformations in rural life. This emerging literature on American farm women encouraged me to consider the farm family economy in all of its dimensions and to follow women carefully as they rushed between kitchens, barnyards, and fields.[6]

Oral histories remind us that people's lives rarely follow the categories that historians cultivate like neat rows, and the interviews suggested that portions of the story of southern agriculture needed to be retold. Scholars, for example, have tended to dichotomize southern farmers and focused on the poorest of the poor, sharecropping families who owned few productive resources other than their strong backs and nimble fingers and received a portion of the crop as payment, or on planter-landlords who managed other people's labor. Many of the people who shared their stories with me, however, belonged to families who owned and operated their farms or to families who rented land and brought to the bargain some tools, implements, and draft animals. To be sure, the proportion of southern farmers who experienced "land-orphanage" grew inexorably during the early twentieth century, with tenants reaching 53.5 percent of the total by 1935, and the burden of tenancy always fell on African American farmers more heavily than on whites. But not all tenants were the same: some paid cash rent, others achieved the status of "standing renters" who paid a set portion of crops for the privilege of farming, and others occupied the bottom rung of the agricultural ladder reserved for sharecroppers. A mosaic of tenure arrangements characterized southern agriculture, and farmers might belong to more than one group simultaneously. The distribution among the various categories varied from place to place across the region. In Mississippi, where the cotton plantation economy of the Delta pulled some 72 percent of the state's farmers into tenantry of one sort or another by 1930, sharecroppers represented more than 40 percent of all farmers. In North Carolina, where owners were a slim majority of farmers in 1930, sharecroppers accounted for about one-quarter of all farmers and one-half of all tenants. Historians have usually concentrated on the end of the spectrum, where planters reigned over sharecroppers, but the interviews direct attention to the farmers who occupied the middle of the tenure continuum.[7]

Descriptions of women's work demonstrated that even as farm families placed their bets on cash crops, many remained as self-provisioning as possible. Subsistence strategies persisted alongside an accommodation to the market. Women continued to be diversified farmers, growing and producing much of what their families ate and putting a "live-at-home" philosophy into practice, even as they participated in production for market.

Narrators often used a common phrase to express this orientation toward subsistence. As people remembered the lean years of the 1920s and 1930s, a standard of well-being that they cited time and again was that their families never missed a meal. A. C. Griffin's experience was typical. Born in 1908, he grew up in a white family that raised cotton on an eighty-acre farm in northeastern North Carolina. Although cash was scarce throughout his childhood and early adult life, he recalled, "We didn't go hungry." As Griffin marveled at his father's ability to provide even that measure of security for twelve children, his wife, Grace, reminded him that his mother "was responsible for a lot of it. She was a hard-working woman." "She *was* a hard-working woman," Griffin agreed, "a very hard-working woman." Not far away in Edgecombe County, North Carolina, Bessie Smith Pender was born in 1919. Her African American family raised cotton and peanuts on rented land, sharing 40 percent of the profits with the owner until they saved enough money to buy a

Lurline Stokes Murray in her yard, Florence, S.C., June 1987. (Photo by Eric Long, 87-16469-9, Office of Imaging, Printing, and Photographic Services, Smithsonian Institution, Washington, D.C.)

farm of their own. "They were tough days," she remembered, but "we were happy, because we had a-plenty to eat, growed plenty of food in the garden. And raised plenty of hogs."[8]

Lurline Stokes Murray, who was also born just after World War I, grew up on the thirty-acre farm her family owned in Florence County, South Carolina, where they raised cotton, tobacco, and corn. She recalled her father's admonition: "'Make something to eat if you don't have nothing else, because a hungry man will steal.'" As a consequence, Murray's mother raised a good garden and kept chickens and cows, and so did Lurline when she had a home of her own. "In all these years," she boasted, "this woman has never been hungry." But, she added, "there's many, many years I didn't know what a piece of money was."[9]

To achieve a goal so basic as never going hungry meant that all family members exploited their own labor, but none worked harder than women. Like so many women of her age and class, Lurline Murray took hard work for granted from the time she was a child. One of her earliest memories evoked a day when at the age of four she stoked a fire, stood on a chair so she could reach the top of the wood stove, and fried a "hoe cake" while her mother chopped cotton on the family's farm. When she arrived at the house, Julia Stokes scolded her daughter for undertaking such a dangerous task alone, but conceded that the cornbread was tasty. "I was trying to help her," Murray explained. "So, I've worked all my life."[10]

The nature of women's work evolved as they combined the duties of daughter, wife, mother, farmer, and friend. Their work varied, too, according to the family's class and race and its stage in the life cycle. Only the wives and daughters of the most prosperous farmers escaped fieldwork, and poor black and white women devoted the most time to crops. In families that could not afford to hire sufficient labor, women's unpaid work in the fields was crucial. As they reproduced the family labor force, adult women might spend less time on fieldwork as children took their place.[11]

Women's work differed, too, according to the cash crops grown. Regardless of the commodity, women performed the most labor-intensive tasks and accomplished them with hand tools. On cotton farms, men broke the land with mules in late winter in preparation for spring planting. As seeds sprouted, women went to the fields with sharp hoes to thin the young plants and returned throughout the growing season to rid rows of grass and weeds that cultivators had missed. By early July, cotton farmers enjoyed a respite from the fields during lay-by time, when all that they could do was hope for good weather, no boll weevils, and high prices. Once the mature cotton bolls broke open and the white lint began to spill from prickly burrs in late summer, harvest began. All members of the family rushed to the fields with long sacks slung around their bodies, dragging behind them in the dirt and growing heavy as pickers filled them with cotton. Backs bent from constant stooping, arms and hands moving as fast as possible, fingers bleeding from frequent encounters with barbed burrs, pickers toiled from first light until dark during a harvest season that might stretch from the sweltering heat of August until the frosty cold of December.[12]

On tobacco farms, the crop dictated a different rhythm and pace. In the spring, women helped transplant fragile seedlings from plant beds to fields. As the plants matured, women hoed grass and weeds left behind after men and boys ran mule-drawn plows up and down the rows. Pulling suckers from along the stalks and flowers from their tops directed the plants' energy to the best leaves. Women's labor became especially vital during July and August harvests. Men and boys usually gathered the mature leaves from the stalks during repeated trips through the fields, but women earned reputations as expert stringers who tied the tobacco to poles that agile boys hung in curing barns. Carefully calibrated and monitored heat turned the leaves golden. Once the tobacco was removed from the barn to the pack house, women graded the leaves according to quality and prepared them for market by tying them into bundles known as "hands." Families took pride in the appearance of their tobacco, and because its presentation influenced the price, tobacco-farming women and men often bragged that the leaves looked like they had been starched and ironed by the time they arrived on the warehouse floor.[13]

Narrators described mothers who wrapped gardening, housekeeping, and child rearing around duties in the field, and many daughters followed in their footsteps. Arnalee Winski's parents sharecropped cotton in south Georgia, and she was the youngest of fourteen children. Her depiction of her mother's days was typical for tenant women. "People had to work back then, honey. We had to work to live. I've heard my mother say that she'd get out and help my daddy plow all day in the field and maybe have two or three little ones. Then when she'd come in, after she got her supper cooked and dishes washed, that she'd do a big washing or mop her house all over and do all that after she got her little ones to bed." On the south Georgia farm where his family raised tobacco and cotton, W. J. Bennett's mother "would go the field of a morning and work until time to fix dinner. Come out long enough to fix dinner and then right back until that evening, just her and us young'uns. She always told us that she might not could learn us to love work, but she was going to learn us how to work. And she did. She didn't have no time off. If she weren't in the field, she was busy at the house. Whenever it come time for a meal, whatever she was doing she quit and went and fixed it, then went back to whatever she was doing." On her family's North Carolina mountain farm, Fredda Davis explained, "Mother's big job was in the house" and "Daddy's work was out in the fields." But "come corn hoeing time, Mama and the children that was big enough was out there in the field hoeing corn. Come haymaking time, okay, Mama'd get out there and she'd take her fork and help build that hay into shocks. . . . And so, yes, [Mama and Daddy's work was] divided, but still so many times that Mama got out in the fields and helped to do what she could."[14]

The peanuts and cotton that Bessie Smith Pender's family raised during the 1920s and 1930s took the effort of all family members. Although her mother routinely regarded field work as part of women's work, directing the family's labor was men's work left to her father. But Pender remembered a particular occasion when gender roles changed and her mother became "the daddy." Her father was too sick to tackle the harvest task at hand—shaking the dirt from recently dug peanut vines in preparation for stacking them on poles to dry and await picking by machine. Her mother gathered the children old enough to help and hauled them to the field in a mule-drawn wagon, leaving her husband at home with the youngest children. Pender, who was about nine years old at the time, considered her mother "the daddy that day. She managed the business. We worked and shaked those peas and come on back that night. When we got back, my daddy had cooked supper for us. The brothers laughed because his biscuits was much bigger than my mother made. The children, they loved to pick and tease. They'd get down and they'd say, 'Come on, get your daddy biscuits. Come on and get daddy biscuits, get your daddy biscuits.' They would just laugh, but they ate those biscuits. They were good, but he just had 'em real fat, like rolls." That the biscuits Pender's father baked were a source of amusement among his sons suggests that for men to do housework was rarer than for women to do fieldwork. Usually, Pender's mother "would just get out there and work and care for us, too. I mean, this is how we done."[15]

Stories about female-headed families reveal how stretched out adult women could be as they tried to meet the demands of farm and home. Virgie St. John Redmond's father abandoned the family in the early 1920s after he went to jail for moonshining. He escaped from a chain gain and disappeared. Determined to stay on the land, Virgie's mother raised four children as a sharecropper in the North Carolina Piedmont. Josie St. John grubbed with a mattock and guided a mule to clear new grounds of trees and stumps for Iredell County landlords, grew cotton, kept a garden, and nursed sick neighbors and washed clothes for wages in order to make ends meet.

A detail from our interview speaks volumes about the demands on a poor white family. Even sunset brought little reprieve for a sharecropper mother and her daughter. Usually, cotton pickers plucked the cotton from the boll in the field. But Josie St. John sometimes made cotton picking a two-step process. By snapping the entire boll from the stalk and then completing the more tedious, time-consuming task of removing the cotton from the bolls indoors at night, the family could increase its productivity, but at the cost of an exhaustingly lengthened work day. "I know when I was growing up," Virgie Redmond said, "I'd get so sleepy at night. I'd get *so* sleepy. But she'd make us pick cotton till nine o'clock."

Whatever the season, the round of work rarely let up. During the summer when the family came to the house at midday for dinner, Virgie Redmond recalled that her mother would insist, " 'Now, kids, get them hoes. We're gonna hoe the sweet taters in the garden while we rest.' That's how we rested." A Fourth of July swim in a creek provided a sweet but brief release from work. On at least one occasion Josie St. John's children mounted a labor protest when they felt that their mother had pushed them too far. Their goal was to gather some muscadine grapes that had ripened in the September sun; the impediment was their fall chore, stripping dried blades from corn stalks and tying them into bundles of fodder for cow feed. "Mama was good to us," Virgie Redmond explained, "she was real good to us, but she didn't want to give us time to get them muscadines. So we just all decided we'd eat us some poison oak and then we'd get sick and then we'd get us some muscadines. We eat the poison oak and it didn't hurt us a bit, and we just had to pull fodder on."

Josie St. John demanded no more labor from her children than she demanded from herself. In hindsight, Virgie Redmond could express only gratitude for a mother who had taught her the value and necessity of work. Time

Virgie St. John Redmond and "Mother resting from all her labor," Grassy Knob Baptist Church, Iredell County, N.C., April 1990. (Photo by Laurie Minor-Penland, 90-4912-26A, Office of Imaging, Printing, and Photographic Services, Smithsonian Institution, Washington, D.C.)

and again, her narrative returned to a familiar refrain. "Mama learned us to work," she said, "that's what she done. She learned us to work." Like her mother, Virgie wanted to farm, and marriage to Mott Redmond in 1937 granted the wish. Before long, the couple bought forty-two acres of land from a relative. As a young mother, Virgie Redmond took her babies to the fields and let them entertain themselves by turning empty oatmeal boxes into toys. Although the four Redmond children worked on the farm, Virgie was determined for them to receive the education that the demands of tenant farming denied her after the seventh grade. She vowed that she would go to any length to keep them in school—even "if I had to wrap up my feet"—and she made good on her promise.[16]

Women invented creative ways to complete all of their household and farm labor. On tobacco farms, for instance, the heat from the curing fire did double duty. Lurline Murray's mother "would kill a chicken and we'd take it down to the tobacco barn and fry it. And we'd put roast sweet potatoes down in the ashes. Mama'd cook cornbread." Murray and her mother also pressed clothes at the tobacco barn, heating their heavy flatirons on wood-fired furnaces. In eastern North Carolina, Nellie Stancil Langley's mother used the heat to boil water in iron pots for canning tomatoes and peaches. "Just put them in the tobacco barn," Langley recalled, "and let them cook while the tobacco was curing."[17] The image of women canning, cooking, and ironing in barns neatly captures the interdependence of farm and household economies and of women's resourcefulness.

In the context of commercial agriculture, women's work on behalf of the household economy was at once more valuable and less valuable than when families pursued semi-subsistence farming. The food that women raised, for example, provided a margin of safety that cushioned families when commodity prices fell. Yet subsistence needs often competed with cash crops for scarce resources, and the interdependence of farm and home did not necessarily translate into adult men's appreciation of women's unpaid domestic labor. Roy Taylor remembered his mother's struggle to secure the supplies she needed to turn apples, peaches, tomatoes, corn, and butterbeans into canned preserves and vegetables that would tide the family over during the winter. Obtaining a hundred-pound bag of sugar was the first of several negotiations. "It's the same story every year," recalled Taylor, the son of tobacco-growing tenants in eastern North Carolina during the 1920s and 1930s. "No money to buy it with. Although the old man and every young'un knows the value of home canning, the old lady has to defend finding the money somewhere to buy sugar for canning and preserving." Once she had secured the vital sweetener, his mother had to nag family members to pick and purchase a bushel or two of peaches at a neighborhood orchard. At season's end, Taylor's mother swelled with pride as she surveyed her handiwork because she knew that husband and children counted on her "to put the vittles on the table three times a day." Nonetheless, she still had "to fuss to get the things necessary to have canned food," Taylor admitted, "and if there isn't plenty of food on the table, they argue at her also. But she has built up a defense against such arguments. She has to do what has to be done, argument or not."[18]

Many farm men and children acknowledged the importance of the work that wives and mothers performed on behalf of their families. A woman reputed to be "a good manager" earned their respect. Although Dorothy Dove's mother hoed and picked cotton on south Georgia farms that the family rented in the 1920s and 1930s, she also canned fruits and vegetables, made jellies from grapes, dried field peas and butterbeans, and pickled cucumbers and watermelon rinds. "In fact, she was a good manager," Dove declared. "There's a lot to that" when accounting for a family's well-being. Walter Anderson's father spared his mother fieldwork on their farm in East Tennessee. His mother, he explained, "was an awful good manager at the house. She was worth more at the house than she was in the field." Clyde Purvis recalled that his mother, too, husbanded the family's limited resources on a south Georgia tenant farm. After his father died in 1922 and left behind a widow and six children, "we were really down," Purvis said. "But we were blessed real good to work. My mother was a good manager, and she knew how to manage, and we'd grow most of what we eat. In the depression days, we got along pretty good because we never had been used to having all these things. So when the depression came, we still grew our own potatoes and meat, and had butter and milk and bread and chickens and eggs. So it really didn't hurt us all that bad."[19]

The agricultural depression of the 1920s and 1930s did, of course, hurt thousands of southern farmers to the point of devastation. By almost any measure—income, investments in buildings and implements, housing, household amenities from running water and indoor plumbing to radios and telephones, diet, and health care—rural southerners of both races fared poorly compared to farmers in the Midwest and Northeast. Contemporary social scientists cataloged the South's symptoms of poverty and economic underdevelopment in painstaking detail. Writers gathered stories and photographers captured images of forlorn women and men, malnourished children toddling on legs bowed by rickets, eroded fields gullied by too much plowing and planting, and ramshackle houses. By 1938, President Franklin

Delano Roosevelt designated the South the nation's "number one economic problem."[20] And yet as Clyde Purvis, Virgie Redmond, Bessie Pender, Lurline Murray, and Nellie Langley testified, many southern farm families stayed afloat, thanks in large measure to the work, managerial acumen, and resourcefulness of women.

Moving the portions of the farm family economy that women controlled to center stage enriches the cast of characters in the story of southern agriculture and complicates its plot lines. Most of the farm women featured here have fallen through the historical cracks. They belonged to a twentieth-century yeomanry—families who owned enough land or tools to control their own labor instead of working for someone else.[21] The material resources they commanded made it possible for women to engage in petty commodity production as well as to guarantee that their families "never went hungry." In turn, women's petty commodity production often made it possible for their families to hold on to what they had. While the South followed a "federal road" to economic development after New Deal farm programs and their progeny began to pump cash into the region in the 1930s, women's economic activities played a large part in ensuring that there was a path to follow or that there were farm families left in the region.[22] Rather than being a vestigial part of the cash-crop economy, the household economy buffered families against the volatility of prices for staple crops, served as an engine of economic change, and reconfigured rural gender relations.

INTERTWINED WITH THE story of southern farm women's economic roles was the story of rural reform. Perhaps no group of American women attracted as much notice from federal and state governments and social reformers in the early twentieth century as did farm women, and yet they have received less scrutiny from historians than other targets of reform such as factory workers.[23] In 1909, when members of President Theodore Roosevelt's Country Life Commission issued the results of their two-year investigation of rural life, they paid special attention to farm women, who appeared overworked and unappreciated, laboring without benefit of new technologies such as electricity and running water that their middle-class city sisters were beginning to enjoy, and deprived of the sociability of town life. Worried that dissatisfied mothers would raise dissatisfied children who would abandon farms for cities, the commission report demanded that "the situation of farm women, upon whom the monotony and loneliness of country life rested most heavily, be given sympathetic attention." As a result, the U.S. Department of Agriculture (USDA) sponsored a special investigation of a select group of farm women around the country to determine their social, labor, domestic, and economic needs.[24]

The federal government translated its concern into policy in 1914, when Congress consolidated various experiments in educating rural adults and youth with passage of the Smith-Lever Act. This legislation established the agricultural extension service—a cooperative effort between the USDA and land-grant colleges—and provided money for farm and home demonstration agents in states and counties that appropriated matching financial support. While male agents sought to modernize the countryside by encouraging scientific agriculture and business practices among farm men and boys, home demonstration agents organized rural women and girls into clubs and encouraged them to apply the principles of science and business to their work as well.[25]

Government focus on farm women is best illustrated by the history of the agricultural extension service in North Carolina, where agents followed a well-prepared trail worn by farm and domestic educators. In 1906, the state's department of agriculture became the first in the South to sponsor separate Farmers' Institutes for Women and to hire female lecturers. During lay-by time in late summer, when crop cultivation had ceased and the harvest awaited, lecturers traveled from courthouse squares to shady groves to instruct farm women in cooking, canning, sanitation, the use of new labor-saving devices, and marketing.[26]

Meanwhile, both black and white agrarian reformers understood that household matters had public consequences. They considered women to be linchpins in the creation of a rural New South of prosperous farms, clean and comfortable homes, healthy children, and vibrant communities. For example, Clarence Hamilton Poe, the ambitious young North Carolina editor whose *Progressive Farmer* was emerging as the premier agricultural periodical in the region, proclaimed in 1913, "The woman, the helpmate and partner of man must be asked to do her part in the mighty tasks of developing here in the South the most splendid type of rural civilization it is possible to work out."[27] Poe's magazine reached thousands of farm homes and through its domestic advice columns urged farm women to fulfill their mission by learning to perform familiar duties in new ways.

Running separate from but parallel to white rural reformers were African American agrarians in the region who viewed farm women as vital to their families on two fronts: they could help families achieve economic independence through farm ownership, and they could refute whites' assumptions of black inferiority, immorality, and shiftlessness with behaviors that signaled thriftiness and respectability. The message of educators at Tuskegee Institute in Alabama spread among devotees throughout the South. While black farm men who attended annual farmers' conferences at Tuskegee resolved to remove the yoke of the mortgage system, to "stop throwing away our time and money on Saturdays by standing around town, drinking, and disgracing ourselves in many other ways," and to dedicate themselves to obtaining better schools, churches, teachers and preachers, black farm women left with a special set of resolutions resounding in their ears. The future physical health and moral welfare of the race rested upon their shoulders; mothers could stem a high death rate among children by feeding and clothing them properly and maintaining clean and well-ventilated homes. Black women could ensure that "strictly moral men and women occupy their pulpits and teach their children." To win the esteem of black men, black women needed to act decorously in the streets, churches, railway stations, and other public places. In addition to talking "in a quiet tone of voice" and refraining from "spitting on the streets," women should dress neatly and modestly, eschewing coarse homespun, red bandanas, and "their hair wrapped in strings" in favor of "neat calico or gingham dresses of modest color" and "the dark colored sailor hat." Finally, women promised to raise poultry and produce for home use and to sell surpluses "to the best advantage" in order to help their husbands purchase a house.[28]

Progressive reformers' increased concern for farm women was but one part of a broader effort to solve the South's social and economic problems.

Men like Poe envisioned a modern South unfettered by tradition and open to change. Diversified agriculture would be an important part of its economic base. Before this goal could be realized, however, farmers would have to share that vision and learn how to increase productivity and income. Education, so the reformers believed, would erode customs that retarded progress and pave the way to a "great rural civilization." Native southerners provided the zeal and passion for uplift, but northern capitalists such as John D. Rockefeller Jr. wrote the checks to support their projects. Beginning in 1902, the Rockefeller-based General Education Board (GEB) pumped millions of dollars into public education and public-health campaigns aimed at farm households.[29]

At the same time, agricultural educators sought ways to reach adult farmers that were more effective than annual institutes. In 1906, also aided by the GEB, agricultural extension agents pioneered a new system of rural adult education. Teacher and entrepreneur Seaman A. Knapp originated the farm demonstration method in Texas while trying to outwit the boll weevil. Farm demonstration techniques depended upon farmers' trying new cultivation practices and witnessing the results. The methods proved so successful that reformers in areas outside the boll weevil district took them up. Then both federal and GEB funds spread the word across the South, arriving in North Carolina in 1908 to complement the farmers' institutes.[30]

With Knapp's encouragement, public schools and state departments of agriculture combined their efforts to reach every member of the farm family. Agricultural agents, for example, targeted farm girls and boys in hopes of influencing them early to adopt modern methods. Corn Clubs for boys promoted new cultivation techniques aimed at convincing participants of scientific agriculture's merits and, according to Knapp, motivated boys to outshine their fathers and shame the men into accepting new practices. Girls' Tomato Clubs likewise taught farm daughters modern gardening and canning methods. In both Corn and Tomato Clubs, agents carefully insisted that children keep all profits in an effort to incorporate them into a market economy.[31]

Tomato Clubs also solved a problem that stumped reformers including Knapp—how to reach farm women with the home economics demonstration method. Knapp used the clubs and educational activities to infiltrate the farm home. "Through the tomato plant," he coached a club agent, "you will get into the home garden and by means of canning you will get into the farmer's kitchen; it will then depend upon your tact, judgment, common sense and devotion to the work as to what you may accomplish for the women and girls of the home."[32] After the creation of the agricultural extension service in 1914, home demonstration agents who organized women into local clubs potentially had the authority to scrutinize and intervene in the most private aspects of family life—from how a woman prepared meals, preserved food, and arranged her kitchen, to how she dressed, raised her children, and kept flies and rodents out of her house. At the meetings of their neighborhood home demonstration clubs, rural women encountered the state, personified in the local agent.

The extension service's programs for women have garnered their share of scholarly criticism. Many students of home demonstration have chided agents for promoting a gender ideology of separate spheres that assigned women responsibility for the home and men responsibility for the farm, ignored the vital part that women played in agricultural production, and promoted women's role as consumers. Critics also fault home agents for favoring women from landowning families who had the material resources with which to accomplish their recommendations.[33]

Close examination of North Carolina's home demonstration program and careful reading of the voluminous reports that agents submitted each year reveal a more nuanced picture of goals and achievements. For example, one of the most striking characteristics of home demonstration is its protean nature. The reports also suggest how official policy and actual practice sometimes worked together and sometimes existed in tension at the local level. While home demonstration represented reform from the top down, its success always rested upon support from the bottom up. Federal and state governments each contributed one-third of an agent's salary, but county commissioners ultimately decided whether to contribute the rest. After a county hired an agent, women themselves were always free to vote with their feet and decide if they wanted to participate in club activities.[34]

Soon after the Smith-Lever Act passed, Jane Simpson McKimmon, North Carolina's first state home agent, traveled from county to county to explain the services that agents offered, to stir enthusiasm among women, and to encourage commissioners to appropriate the local share of an agent's salary. Some counties and some women responded to the message more fervently than others. Enthusiasm for home demonstration grew in 1918, when the extension service accepted the federal government's charge to increase food production and preservation during World War I. As time passed and the program became more popular with rural women, the appointment and retention of agents could become embroiled in local political brouhahas that inspired club members to join the fray. When Alamance County commissioners, for example, considered withdrawing their support for the home agent in 1923, club members "waged quite a battle" on her behalf. In a letter to the editor, a club leader told readers of the *Alamance Gleaner* that she wondered where the money would go if not to an agent's salary. To some political scheme? "If any people ever deserved the little that they are getting," the writer argued, "it is the women in the farm homes" who worked so hard. If the commissioners failed to follow club members' recommendations, she encouraged women to "vote where it will help" and to flex political muscles on behalf of home demonstration work. In the early 1930s, as commissioners in many counties eliminated their support for the salary of home demonstration agents in an effort to balance depression-starved budgets, loyal clubwomen across North Carolina put elected officials on notice that supporting the extension program was key to earning their votes.[35]

Class and race circumscribed the work of home agents. Among white home demonstration members, women from landowning families always outnumbered those from tenant households. Like those in other southern states, North Carolina's extension service gave exceedingly short shrift to African American farm women. Eight years passed after Smith-Lever's approval before the state hired its first permanent black home agents.[36]

Home demonstration agents did often promote cultural ideals that encouraged farm women to mimic aesthetic standards set by middle-class women who lived in towns. Lessons in home decorating and yard design promoted color choices, furniture arrangements, and accessories foreign to vernacular tastes and borrowed planting schemes from landscape architects that were more suitable for cities and suburbs. Repeatedly, home agents emphasized that farm women had achieved an important milestone when in appearance they could not be distinguished from town women. Nonetheless, agents often helped women find a means to new ends with a minimal expenditure of cash that allowed them to "make do" with resources at hand. To improve the appearance of their yards, home demonstration clubwomen scoured the woods to secure native shrubs and flowers that nature provided or exchanged cuttings from favorite plants. Women who entered club-sponsored kitchen improvement contests boasted that for a few dollars they had brightened walls with a coat of paint, made floors easier to clean with a covering of linoleum, acquired new equipment, and arranged the workspace to ease their domestic labors. Agents taught women how to remodel old clothing into more flattering styles and make becoming hats for just pennies.[37]

Home demonstration agents played a crucial role in bringing modern methods of public health into farm homes and communities. During the early 1920s, agents began to cooperate with public-health nurses and doctors in their attempt to educate women about prenatal care, child nutrition, and health. As the number of pellagra sufferers in North Carolina grew at an alarming rate, home agents became the eyes and ears of the state board of health. They identified victims whose blotchy, red skin betrayed the fact that their diets lacked necessary amounts of niacin. Agents and club members then taught farm women how to prevent the disease and helped them obtain the foods that could cure it. In the 1930s, agents augmented the state's relief programs, helping clients grow and can vegetables and fruits. In the process, some agents reached a new understanding of the dire conditions on the poorest farms.[38]

In North Carolina and throughout the South, women's farm production was a primary focus of home agents' programs. From the outset, agents encouraged women to sell the products of their labors and sought to enhance the value of their earnings. The first objective of home demonstration work in the South, leaders at the USDA proclaimed in 1917, was "to develop a skill that shall result in economic independence of girls and women in the country." Home demonstration leaders recognized that the country home represented a "producing as well as consuming center" that contributed "to the income of the farmer" and often measured "the difference between profitable and unprofitable farming." Increasing farm incomes was the first step toward raising a family's standard of living and fostering community improvement projects.[39] Home demonstration agents organized cooperative marketing of farm women's goods, and in the 1920s they inaugurated curb markets that linked club members with buyers in town.

In addition, home agents recognized the value of women's unpaid labor on behalf of their families. They assigned a monetary value to the vegetables that women canned and the clothes they stitched. Agents thus prevented a "pastoralization" of women's domestic work that made it appear not to be work at all.[40]

When considering the activities of home agents, it is useful to take a page from literary critics who argue that an author is not the sole creator of a text; rather, the text is a joint creation of writers and readers. In much the same way, agents and club members collaborated in writing the texts of lessons when they came together at club meetings. To foster interest and retain members, agents consulted with clubwomen as they set annual agendas of lessons to be studied. Clubs could concentrate on gardening, food preservation, sewing, home improvement, marketing, and other topics from year to year. Although agents' annual reports read like conversion narratives in which skeptical women become true believers in the new methods shown, there are enough signs of doubt to indicate that members adopted methods that suited them and jettisoned the rest. Reading between the lines, it is clear that many women with years of domestic experience mocked an agent's canning techniques or the way she cut out a dress and turned up their noses at the suggestion that cabbage steamed for a few minutes tasted as good as pork-seasoned greens boiled all day.[41]

On the other hand, many club members welcomed the opportunity to continue their educations, to learn new ways of completing and managing household tasks, and to obtain information from so-called experts. They appreciated information about what constituted healthier diets so they could feed their families more nutritious foods and could pack lunches that would improve their children's capacity to learn at school. They enjoyed planting a greater variety of vegetables in their gardens after receiving recommendations from agents. They testified that better-arranged kitchens saved time and labor. They valued advice that made poultry flocks more profitable and inspired new ideas for reducing expenses and earning money.[42]

Loyal club members counted the reprieve from lonely labors a chief attraction of monthly meetings. In the late 1920s, clubwomen in Johnston County described the regular visits of their home agent as "the greatest event in the lives of the women" in rural communities. "Many hard-working, tired mothers look forward with so much pleasure to the day 'our club' meets as the gala day in their lives," wrote the recording secretary of the Cleveland Club. At club gatherings, women conquered shyness and acquired confidence as they learned how to conduct meetings. They assumed leadership positions at the local and county levels that served as stepping-stones to participation in the statewide federation of home demonstration clubs that formed in 1924.[43]

Certainly, farm women were free to accept or reject home demonstration. In North Carolina, thousands flocked to club meetings. Membership among white women grew from some 2,800 in thirty counties in 1916 to some 20,600 in sixty-six counties in 1932; in 1922, some 1,900 black women belonged to clubs in six counties, and a decade later the number had grown to more than 3,100 in seven counties. In 1935, the extension service boasted that black and white home agents had reached more than 54,000 women and girls with their lessons that year. Home demonstration was the largest women's voluntary association in the state.[44]

THIS BOOK FOLLOWS farm women as they secured resources for themselves and their families by taking advantage of expanding markets and government-sponsored expertise in the rural New South. Its geographical and evidentiary heart lies in North Carolina and other South Atlantic states. Chapter 1 examines women's dealings with itinerant merchants who brought an array of manufactured goods to their front porches and back doors. Peddlers, hucksters, operators of "rolling stores," and sales agents known as "Rawleigh men" and "Watkins men" traveled the countryside, transformed farm households into places of consumption, and sometimes turned hierarchies of gender and race on their head. Frequent visitors to farm homes, these itinerant merchants had become as invisible in the telling of the rural South's economic history—where country stores commanded the spotlight —as had the women with whom they dealt.

Chapter 2 turns to farm women's production for market. Women's "butter and egg trade" defined rural economic life day in and day out and, often unbeknownst to the sellers, connected them to an economy that extended far beyond their neighborhoods and intersected with regional and national networks. While this decentralized economy is hard to trace, oral history evidence provided the opening wedge. Considered in isolation, swapping eggs for groceries might appear incidental and pale in comparison to sales of cotton or tobacco in a good year. But taken together and placed within a broader context, such routine transactions added up to more than survival. The cash and credit that women accumulated were central to the family farm economy, and women often used their income to leverage some independence from their men folks.

Chapter 3 concentrates on a transitional period in poultry production during the first four decades of the twentieth century. Women took advantage of growing demand for eggs and fowls, new ways to produce them, and new ways to get them to market. Innovative women showed that poultry paid and laid the foundation for the agribusiness poultry industry that emerged after World War II.

The next two chapters shift the focus from farm women to the home demonstration agents who worked with them. Chapter 4 describes the professional culture that developed among white home agents in North Carolina as they acted upon new job opportunities that the agricultural extension service created. Just as farm women combined old and new ways of buying and selling, women who pursued professional duties struggled to reconcile new possibilities and old ideas about women's proper roles. Chapter 5 analyzes the work of African American home agents and stresses their achievements as they offered their constituents a link to valuable government resources. Despite the racism that hobbled their efforts, black agents gave new meanings to traditions of self-help, mutual aid, and racial uplift.

Chapter 6 serves as a coda to the book. It weaves together several themes by considering the changing meanings of one of the most prosaic items familiar to farm women, the feed bag. Seemingly simple symbols of women's ability to "make do" with precious little, feed bags became complicated symbols of market production and mass consumption.

This book and the women whose stories drive its narrative invite us to take another look at the southern countryside. By foregrounding women's engagement in rural commerce, we can see women participating in the so-called incorporation of America on their own terms.[45] As buyers and sellers, farm women shaped increasingly complex economic, social, and cultural networks even as they were influenced by larger processes at work all around them. By reconsidering the work of home demonstration agents and clubwomen, we see that when viewed from state and local angles, government-sponsored rural reform assumed a more contested and less coercive character than when viewed from federal and regional perspectives.

STUDENTS OF AMERICAN women's history often encounter southern farm women through the story of Mollie Goodwin, one of the mothers of the South that Margaret Jarman Hagood interviewed in the 1930s. It would be easy for readers of the popular textbook in which "Of the Tenant Child as Mother to the Woman" is anthologized to dwell on the pathos of Mollie's story and miss the note of optimism that she sounds at the end.[46] The case study begins when ten-year-old Mollie is already captive to the patriarchal system of labor and fertility that determines the lives of poor rural women. Because she has to help her mother with weekly washings, Mollie cannot attend school regularly; rather than reciting lessons, she remains at home and inhales the stench of soiled baby diapers as she scrubs. At the age of eleven, Mollie menstruates for the first time on the very night that her mother gives birth to her sixth child.

One of the most poignant moments of the vignette comes when a light-hearted Mollie, now fourteen, joins her family for a rare outing on the Fourth of July. She wears a new dress made of rose and green taffeta, a cherished gift from her father in appreciation for her hard work. Mollie's high hopes collapse, however, as soon as she arrives in town and jumps off the wagon. She has started her period early and menstrual blood spoils her dress. To hide the embarrassing stain, she sits numbly at the base of a tree the entire day. All about her, folks celebrate the country's independence while Mollie is a symbolic captive to biology.

Eager to leave the farm, teenaged Mollie finds a job in a tobacco factory. She likes the company of other working girls and enjoys spending her wages on new clothes and shoes. At the urging of her father, however, Mollie returns to the farm, lured by his promise that she can board in town with an aunt and attend high school. But before those plans materialize, Mollie's mother and brother take sick and need her care. By the time her mother delivers her ninth child, Mollie has lost interest in continuing her education.

Mollie appears destined to follow in her mother's footsteps. She marries Jim Goodwin, a tenant farmer ten years her senior. Within five years, she is pregnant for the third time. Although her first two children, ungainly and sickly, have disappointed Mollie, the third promises to be lively because it is "the kickingest baby you ever felt." Suddenly, the baby stops moving inside her, and by the time of delivery the fetus has disintegrated.

Complications following the miscarriage interrupt her fertility, and a decade passes before Mollie, now thirty-three, has another child. The little girl is the apple of her mother's eye. How was Mollie Goodwin trying to guarantee that this daughter would enjoy more options than she had, that what she passed on was not simply a replication of her own position? The doting mother, Hagood noted, "began selling eggs to bring in a little money all during the year so she could buy cloth to make her daughter pretty clothes." The eggs Mollie sells become a symbol of hope for a better life for her daughter and a means of securing it. Like Mollie Goodwin, the women in this book looked beyond despair and used the resources at hand to shape their own lives and the larger world around them.

CHAPTER ONE

Rolling Stores

FEW STRANGERS CROSSED the hardscrabble landscape that Harry Crews evoked in his memoir of childhood in south Georgia during the 1930s and 1940s. Crews remembered that the Jewish peddler, driving a team of mismatched mules, "came into [his] little closed world smelling of strangeness and far places." Bolts of cloth, needles, thimbles, spools of thread, forks, knives and spoons, and a grinding stone that "could sharpen anything," staples, nails, mule harnesses, "and a thousand other things" filled the inside of the peddler's wagon. Frying pans, boiling pots, washtubs, and mason fruit jars festooned the outside.[1]

It was Harry's mother, Myrtice Crews, who negotiated for these treasures. The peddler "did business almost exclusively with women," Crews recollected, "and whatever they needed, they could always find in the Jew's wagon. If they didn't have the money to pay for what they needed, he would trade for eggs or chickens or cured meat or canned vegetables and berries." To arrive at a price, the peddler and Crews's mother performed a ritual of bargaining. No matter how much Myrtice Crews protested "I ain't got the money," she often succumbed to the peddler's invitation to feel a bolt of cloth; and, after more haggling, he might offer to take corn and hay to feed his mules in payment for the fabric. Harry's mother signaled the end of the transactions when she "silently took two brown chicken eggs out of her apron and gave them to him" in exchange for peppermint balls. "With the candies melting on our tongues," Crews wrote, "we stood and watched him go, feeling as though we had ourselves just been on a long trip, a trip to the world we knew was out there but had never seen," a trip financed with the products of his mother's labor.[2]

Modest negotiations like those conducted by Myrtice Crews in her backyard occurred routinely during the first half of the twentieth century. Ubiquitous but elusive, transactions with itinerant merchants have been overlooked by scholars who assume that country stores took center stage in the rural New South. Analyzing stories about itinerant commerce rather than treating them as quaint anecdotes enriches the understanding of the New South's consumer economy and broadens the cast of characters who shaped it. Recent studies demonstrate that southern country stores were economic institutions and social spaces where white men held the upper hand.[3] Oral history interviews, reminiscences, regional fiction, and trade literature illuminate a universe of buying and selling where itinerant merchants prevailed, where the farm household was a place of consumption as well as production, and where male dominance and white supremacy might be challenged.

A consideration of itinerant commerce joins a lively historical literature that charts the complex story of how consumer culture spread across the United States. Once preoccupied with middle-class shoppers in the urban Northeast, scholars have discovered that the story's subplots vary by region, race, gender, and class, revolve around the agency of buyers as well as the power of businesses, and unfold at different times in different places. This look at itinerant commerce focuses less on cultural historians' questions about theories of consumerism and the meanings of goods than on the concerns of social historians—the nature of buying and selling, the power relationships that defined those transactions, and the context in which they occurred. It highlights human relationships rather then anonymous representatives of the market.[4]

Several kinds of mobile merchants operated in the rural South well into the twentieth century. Some, like the Jewish peddler of Harry Crews's youth, were independent entrepreneurs who sold from their packs or wagons. Other itinerants operated what were popularly known as "rolling stores," portable extensions of permanent businesses. Still other "traveling men" represented large manufacturing firms that offered a diverse line of goods and provided elaborate advice on sales techniques. Two firms whose retailers frequented rural southern homes were the W. T. Rawleigh Company and the J. R. Watkins Medical Company, both headquartered in the upper Midwest but boasting factories and distribution centers all over the United States, with several strategically located to take advantage of southern markets.[5]

Itinerant merchants extended the world of manufactured goods into the countryside. Although they certainly did business with any willing customer, they appealed to women and African American buyers, who entertained fewer options as consumers than did white men. Itinerant sellers accommodated the needs of black and white farm women, whose patronage of town and country stores was often limited by constraints on their travel and by their discomfort in commercial places where men gathered to do business and visit. Because itinerant merchants took the products of women's labor in trade, women customers could combine domestic production with consumption in a mix that suited their needs and pocketbooks. Contrary to the assumptions of cultural historians that sellers always had the upper hand, these customers were shrewd buyers rather than easy dupes, and they enjoyed the challenge of bargaining with traders. Besides the goods that filled a peddler's pack or a sales agent's sample case, itinerant merchants arrived with news from beyond the immediate neighborhood, a commodity prized by women who kept close to home. Furthermore, African American men found doing business with mobile merchants attractive because what and where itinerants sold provided alternatives to trading with southern country storekeepers whose racist assumptions shaped access to credit and goods.

Itinerant merchants in the New South joined a long line of mobile sellers that began with peddlers. American business practice and lore usually associate peddlers with antebellum Yankee entrepreneurs who sold clocks, tinware, patent medicines, cloth, and notions to country people. In the early nineteenth century, peddlers served an economic purpose as distributors for small-scale manufacturers and importers. But as historian Jackson Lears has argued, they also were the heralds of a nascent consumer culture and touchstones for the anxiety and ambivalence that the spread of a market economy generated. Here today and gone tomorrow, itinerants crossed the boundaries between local neighborhoods and the cosmopolitan world; they represented the "magic" of the marketplace, with its mixture of danger and allure. Peddlers inspired a rich folklore and humor that mirrored the apprehensions that accompanied the quickened pace of nineteenth-century commerce. In literary representations and the popular imagination, peddlers aroused suspicion as outsiders and clever tricksters who supposedly found women, in particular, to be vulnerable to their verbally seductive sales pitches.[6]

As peddlers extended the market's reach into the antebellum South, anxieties about both race and gender aroused suspicions. Newspaper editors and storekeepers accused peddlers of "cheating women and children."[7] At least one husband agreed. In 1826, a Georgia man accused peddlers of convincing his young bride "to buy a considerable quantity of store-goods" without his "knowledge or approbation" and was so incensed that he took out a newspaper advertisement warning vendors away from his house for eighteen months.[8] Peddlers also stirred distrust among slaveholders. Planters feared that itinerants trafficked with slaves who offered pilfered livestock in trade and worried that peddlers were abolitionists in disguise sowing the seeds of rebellion among bondmen. In the wake of insurrection panics, slaveholders cast skeptical eyes on itinerant merchants who were strangers in their midst. After John Brown's 1859 raid at Harpers Ferry, for example, whites in several Georgia towns questioned and harassed traveling salesmen from the North, who were thought to sympathize with the abolitionist. An idle remark was enough to provoke a committee of white citizens in Columbus, Georgia, to scrutinize the "many persons from the free States now traveling through our neighborhood for the ostensible purpose of selling books, maps, rat traps, etc.," and such harassment inspired the salesmen to catch the next train or steamship home.[9]

What those who study peddlers have ignored—but is implicit in their accounts—is that the harshest critics of itinerants were white men, who constructed unsavory images of peddlers and their female and African American clientele in an effort to control access to the marketplace. Centering the analysis of consumption around gender and consumers' perspectives, however, places peddlers in a different light. Perhaps peddlers posed a danger because they took their wares to buyers who were not male heads of households. By extending the market, peddlers and their patrons undercut the domestic authority of men. From the vantage point of women and slaves, peddlers were welcomed visitors. The same was true for their postbellum descendants.[10]

Although the popular literature of the New South perpetuated images of peddlers as duplicitous tricksters, these portrayals camouflage the precarious social conditions of their work. After the Civil War, recent newcomers from Eastern Europe got a foothold in business by peddling in the South. With little capital to invest, immigrant peddlers often catered to poor whites and African Americans, economically marginal figures like themselves. Stereotyped as shady dealers, peddlers were in reality vulnerable salesmen who were easy targets of criminals as they traveled alone hauling merchandise and carrying money. Such was the fate of Francis Brice, an Irish peddler who sold his wares in southeastern North Carolina. In 1878 as Brice passed through a remote marsh, two local men shot him, wrenched a pistol from his hand, and clubbed him with his own walking stick. After stealing Brice's money, the robbers left him to die. Similar bad fortune befell Samuel Tucker, a Jewish peddler from Richmond, Virginia, whose route traversed the North Carolina Piedmont. During an 1892 selling trip, Tucker sought overnight lodgings in Franklin County and was robbed and murdered where he rested. The culprits tossed his body into a vine-choked ravine, where it went undetected for months.[11] As these cases suggest, representing peddlers as tricksters endowed them with more social power than they actually wielded.

Popular mythology also distorts other characteristics of peddlers' work. Like the itinerant whom Harry Crews remembered, many peddlers became familiar outsiders rather than feared strangers. Though customers surely did not count them as neighbors and frequently associated immigrants with exotic, foreign places, between spring and fall of each year peddlers often traveled predictable circuits that brought them back to the same customers month after month. For example, in the early twentieth century a man known only as "Peddler Black" was a regular visitor in Owen County, Kentucky. While residents later debated his ethnicity and pondered his "mysterious origins," at least one woman asserted that "he *was not a stranger to us*."[12]

Although general stores had sprouted at nearly every dusty crossroad by the 1880s and 1890s, itinerant merchants continued to find a niche in the New South. Even as railroads connected the region to national markets and the region crawled with drummers who sold wholesale to storekeepers, access to the manufactured goods that expanded briskly after the Civil War varied according to place, race, class, and gender. Members of the new middle class who lived in the South's burgeoning towns and cities could patronize stores and specialty shops that their country cousins could only imagine. Country stores offered lines of merchandise that ranged from patent medicines to plow points, but peddlers and other sales agents brought their wares right to the homes of customers for whom the rural retail outlets were alien places.[13]

General stores, as historian Ted Ownby has noted, "were not settings for racial equality." Nor were they settings for gender equality. Black and white men struck bargains with merchants who provided credit for farm supplies, groceries, and dry goods in exchange for liens on their crops. White men gathered around stoves to play checkers or loafed on porches, passing the time with ribald humor and neighborhood gossip, the conversations sometimes lubricated with whiskey. Although white storekeepers welcomed African American men as customers, they kept a tight fist on credit, monitored their purchases, directed them toward inferior goods, and did little to ease the tensions that black men felt as "conspicuous outsiders." In the era of Jim Crow, the stores became stages where southerners choreographed steps in a new dance of race relations. Sometimes black and white men mixed freely and without conflict, but all too often African Americans suffered verbal epithets or physical jostling that diminished any joy attached to buying.[14]

Bastions of male customs and habits, country stores were not particularly hospitable places for women to shop, either. In his classic study of southern country stores, Thomas D. Clark noted that women approached these centers of commerce with reluctance and uneasiness. "They stood near the

Men at filling station and general store, 1939. (N85.2.25, Archives and Records Section, N.C. Division of Archives and History, Raleigh)

front door with embarrassed grins on their faces showing clearly mixed feelings of eager curiosity and shocked modesty," Clark wrote. "They were caught in the unhappy situation of not knowing whether to stay until someone came to serve them or to leave the store." In the 1930s and 1940s when photographers from the Farm Security Administration (FSA) traveled the back roads of the South, they found men lounging on store porches and gathered inside; women were hardly to be found. The woman who did venture across a store's threshold might, in William Faulkner's words, have to pass "the squatting men . . . spitting across the heelgnawed porch" and training their "ranked battery of maneyes" on her.[15]

Practical constraints joined ideology to restrict rural women's visits to stores. Preoccupied with bearing and rearing children much of their adult lives, rural women simply had less chance to get away from home. Basic house-hold maintenance filled their time. When Sara Brooks was growing up in the Alabama Black Belt during the 1920s, her mother rarely accompanied her father to town on Saturdays. Brooks's mother reserved the day for cleaning the house, sweeping the yard, and cooking Sunday dinner.[16]

Although the products of women's labor often provided the currency for trade, men usually did the shopping. Sara Brooks reported that before her father went to the store, he asked his wife, "'Now what to get?' [and] [s]he would write him a little list and give it to him." In eastern North Carolina, A. C. and Grace Griffin established a similar pattern in the 1930s and maintained it for years. "When we first got married we didn't have any transportation the first five years," Grace Griffin explained. "On Saturday morning [A. C.] and his daddy would go to town in the mule and cart and buy the week's groceries. Then we had a young couple . . . move close to us. . . . He and A. C. would go to town together on Saturday morning a lot of times and each one would carry whatever eggs we had left over from the hens that week and buy the groceries. Well, the pattern was set by the time we got an automobile and I learned how to drive it."[17]

The shopping arrangement that the Griffins worked out also followed from expectations about proper public behavior, which differed for men and women. When women went to town to buy groceries, Grace Griffin had been raised to believe, they should dress up, whereas men were free to wear their work clothes away from home. If A. C. did the shopping, she was spared the trouble of outfitting herself for a "public" appearance.[18]

Itinerant merchants understood these facts of economic and social life and turned farm households into sites of consumption as well as production. In reminiscences of rural life in the early-twentieth-century South, peddlers remained familiar figures. Filtered through the lens of childhood, recollections ascribed magical qualities to peddlers who inspired fascination and fantasy. The peddler's visit to the Wake County farm where Bernice Kelly Harris grew up in the late nineteenth century remained a vivid memory for the North Carolina writer forty years later. When the peddler opened his packs, Harris recalled, "It was Aladdin's lamp, an adventure in wonderland, a glimpse into the gold-haired prince's palace beside the sea. For the rings and things conjured up a way of life that glittered." The peddler, according to Harris, "was not so much a person to the children as a symbol, like Santa Claus."[19] Harris was not alone in her fond memories. Itinerant sellers "who visited the farms of those days and brightened the lives of those who lived there" figured prominently in Caroline S. Coleman's reminiscences about a turn-of-the-century childhood in the South Carolina Piedmont. The peddlers who "linked the remote countryside with the marts of trade" in her part of the world were usually Italians, Syrians, or Irishmen. Children playing in the yard often spied the peddlers first and relayed the exciting news to their grandmother. As the peddler unfastened his pack, Coleman remembered, "we watched in wide eyed wonder" as he revealed the "riches of Araby."[20]

For children like Harris and Coleman, peddlers represented miraculous gift bearers. But for adult women their visits provided opportunities to bargain for household necessities and a few small luxuries and to embroider their roles as good domestic managers. While young Bernice Kelly enjoyed the aroma of soaps and admired the ribbons, lace, and brooches that the peddler displayed, her mother calculated which goods she could afford to buy and which trinkets she had to pass up. After engaging in a form of window-shopping in her own home, Bernice appreciated that her mother "had spent her butter and egg money to let us enjoy looking at the wares." When the peddler arrived at the home of Caroline Coleman's grandmother, the older woman invited him to turn her parlor into a showroom for the fine shawls, towels, tablecloths, pillow shams, combs, and beads that he pulled from his pack. Word of the peddler's visit traveled the "grapevine telegraph," and soon "every colored woman on the place" came to examine his stock and to join the "fine art" of trading with him. Buying turned into a collective activity, and having several women match bargaining wits with a peddler might have afforded them some advantage. After the dickering had ended, Coleman's grandmother offered to board the peddler for the night, her gesture of hospitality repaid with stories "of his home country in the great world beyond our doors" and with a gift such as fine linen towels.[21]

Hassan Mohamed was a Lebanese peddler who gained a foothold in southern commerce. Soon after he landed in New York City in 1911, Mohamed headed to the Mississippi Delta, where friends helped him get a start.

A Clarksdale, Mississippi, storekeeper "gave him $27 worth of merchandise in [a] little suitcase," and Mohamed set out on foot to sell. A limited command of English and an unfamiliarity with American currency put Mohamed at the mercy of his customers—a dilemma that turned the image of the crafty peddler on its head. "He said he didn't know the five dollars from the ten [dollar bill] or the one," recalled his widow, Ethel Wright Mohamed. "But he said he felt like any house he stopped, they'd buy whatever they bought out of the suitcase, they would make the change and he felt like everyone was honest. He didn't think anybody beat him out of any money."[22]

Black sharecroppers who worked Delta plantations patronized Mohamed, and he counted one woman in particular as a "special" customer. If she and her family were working in the fields when he arrived, Mohamed napped in the shade until she came to the house. "Then he would go up on the porch," Ethel Wright Mohamed explained, "and she would have some of the neighbors come, and they'd buy everything out of his suitcase right there. And he said she would run her hand up under the rug and get the money she had hid in the house. And he thought that was the most wonderful thing he'd ever seen. And so that was her bank . . . and that was his store." Within three years, Mohamed had earned enough to buy a wagon, allowing him to expand the line of goods he carried and the territory he covered. By 1922 he could afford to buy a store in the hamlet of Shaw and to settle permanently.[23]

Trading with peddlers was especially appealing to women. They could weigh their choices in the privacy of their back yards or front porches. They shaped the space where goods were displayed, calibrated their value, and gauged their own capacity to spend. They might even practice their bartering prowess before an appreciative audience of neighbors. On home turf, women incorporated commerce into a domestic economy where they operated from a position of strength and helped define the terms of trade.

The peddlers' lessons were not lost on storeowners. Some rural storekeepers realized that they could tap their market's full potential only if they took their merchandise to women's homes. At these "rolling stores"—trucks or wagons fitted out with merchandise—customers could buy according to their financial means or personal whims. Hal Edmonds's grandfather knew that the farm wives and daughters in the North Carolina mountains made infrequent trips to the country store or to the Madison County towns of Mars Hill and Marshall. But every week O. S. Edmonds took a scaled-down version of his permanent store to customers, trading merchandise for eggs, chickens, and butter. In the 1940s, Hal worked as a clerk for his grandfather. "Most all that traded in the regular store," Hal Edmonds explained, "they was mostly men. Men would come to the store and stay about all day and their wives, whatever they wanted, they would write it down or tell the men. Women very seldom ever come to the store. Like in the rolling [store] business, it was all different. It was all women."[24]

Because Edmonds followed regular weekly routes, women anticipated his arrival and met him at the end of their lanes, ready to offer baskets of eggs or coops of live chickens in trade. "They'd like to come out right beside the house," Edmonds recalled, "sit down beside the road. See, you started up a little road up one of these hollers, and they could hear you coming. Wasn't no traffic." From the shelves and aisle of the specially rigged truck, Edmonds supplied primarily the same merchandise that stocked the shelves of his "regular store." He also took special orders for bulky items, such as a hundred-pound bag of salt during hog-killing season, or custom goods, such as shoes, which he delivered the following week. The rolling store business continued in Madison County until the 1950s, when better roads and more automobiles eased travel and chain grocery stores arrived.[25]

Rolling stores also served rural customers in south Georgia, where Mioma Thompson grew up the daughter of farm wage hands during the 1930s. Her mother patronized stores in the town of Tifton and the rolling stores that stopped at their house. Rolling stores, Thompson remembered, "carried material, flour and sugar, flavoring, or this liniment. A lot of people back then was bad to rub with liniment. Just whatever they could put on that store that would keep, that's what they'd take." By the time the mobile store arrived, "Mama would always save her money and her eggs. Course, we had what we wanted to eat. But about time for [the rolling store] to come, she'd go to saving them eggs up and put her a chicken in the coop, fatten it up. That's the way she got what she needed." Mioma's mother controlled the egg money and used its proceeds to buy staple goods and to indulge a child's sweet tooth when the mobile merchant visited.[26]

"The inside of a rolling store. The average daily sales amount to $60, for which a lot of them are paid in produce from the farms. A tank of kerosene can be seen on the left." (Photo by Marion Post Wolcott; LC-USF 34-51439-D, Coffee County, Ala., April 1939, Farm Security Administration/Office of War Information, Division of Prints and Photographs, Library of Congress, Washington, D.C.)

Buying at the rolling store allowed customers to acquire needed goods and to enjoy a form of entertainment. In East Tennessee, until the mid-1940s, Della Sarten traded butter and eggs with a man who ran a rolling store "that was just packed full." A man from Sevierville "come along with cloth, had anything like they had in a grocery store," Sarten recalled. "He come around once a week. I had eggs. Sometimes I'd sell two or three hens. Then I'd have a list wrote out, and I'd just call it off, and they'd just hand it out, and on they'd go." To the merchant's dismay, not everyone made their choices as quickly as Sarten did, and he recounted to Sarten his experiences with other customers: "My next door neighbor, she'd climb up in their truck, and she'd look at the cloth and she looked at everything for about an hour. He said one day, 'If everybody was like you, we'd get along.' They'd come back in Sevierville way in the night, you know, people taking so long to trade."[27] In contrast to the pragmatic Sarten, her neighbor turned the storekeeper's visit into a leisure activity. Either way, the seller had to keep the customer happy.

A third kind of itinerant merchant was a common visitor to rural southern homes—agents who represented large manufacturing enterprises and specialized in a line of goods. Salesmen of musical instruments, lightning rods, and patent medicines pervaded the autobiographical novels and folk plays of North Carolinian Bernice Kelly Harris. When a stranger appeared in the fictional community of Beulah Ridge, for example, the women speculated that he must be selling farm periodicals or Canadian lace. Like peddlers, these "traveling men" often doubled as entertainers; their skills as raconteurs and musicians complemented their sales pitches, and customers prized their talents, company, and the news they brought with them as much as they prized the merchandise they sold. In Harris's play *Pair of Quilts*, Sudie Johnson rated the traveling men who boarded in her home. "I thought the 'Native Herb' man was real entertainin' with his fast pieces on the fiddle, and the lightnin' rod agent with han'chief tricks," Sudie said, "but the grapherphone man's got 'em all beat with his funny jokes."[28]

Like the members of her own family, the women of Harris's fiction welcomed sales agents, but the men often tried to ban them or banished them outright. One woman of the neighborhood reported that the squire of Beulah Ridge "stopped agents from comin' on his land, told 'em to stay away

"A rolling store which goes from door to door selling groceries, hardware, dry goods, drugs and a variety of household and farm supplies." (Photo by Marion Post Wolcott; LC-USF 33-30398-M2, Montezuma [vicinity], Ga., May 1939, Farm Security Administration/Office of War Information, Division of Prints and Photographs, Library of Congress, Washington, D.C.)

from the Ridge just like he owned it all." In Harris's novel *Purslane*, the patriarch John Fuller so despised solicitors that he "had once put up a sign at the road: 'No agents allowed here.'" The warning had backfired, however, attracting rather than deterring curious salesmen. The "Native Herb" man was among those who ignored the posting and annoyed John Fuller by interrupting his plowing with his sales pitch. But when the agent "mentioned indigestion," John's wife, Dele, "entered the conversation," Harris wrote. "The result was she took the agency for the herb pills," which "became a family and neighborhood standby" for all manner of ailments and also turned a profit for Dele. How frequently women sold medicines is uncertain. In *Moth-*

ers of the South: Portraiture of the White Tenant Farm Woman, sociologist Margaret Jarman Hagood described the pitch of a woman selling salves, tooth powders, and a tonic called Vi-Ava.[29]

These fictional sales agents were in some ways prototypes of the men who sold the products of the Rawleigh and Watkins companies. To this day older rural southerners recollect regular visits by "the Rawleigh man" and "the Watkins man." Although they often put these sellers in the same category as peddlers and rolling-store merchants, salesmen for Rawleigh and Watkins were, in many respects, a breed apart. When they arrived at farmhouses with their sample cases, they were retailers acting on behalf of large companies whose founders had coached them in the latest sales techniques. These itinerant merchants represented a new, intensified phase in the creation of a national market for brand-name consumer goods. But gender, race, and local customs shaped the way southerners encountered the market.[30]

Although Rawleigh and Watkins retailers visited rural southern homes frequently, the parent companies were located in the upper Midwest. Based in northern Illinois, the W. T. Rawleigh Company headquarters was less than two hundred miles from the J. R. Watkins Medical Company hometown of Winona, Minnesota. W. T. Rawleigh began modestly in the late nineteenth century, peddling to farmers the medicines that his first wife—whom he later divorced—manufactured and bottled in their home near Freeport. By the early twentieth century, Rawleigh envisioned the sales potential and the company began building factories in cities across the United States and Canada. The birth and development of the J. R. Watkins Medical Company followed a similar path, starting with a popular liniment in 1868 and expanding its inventory and sales territory in the early twentieth century.[31]

A dizzying array of products rolled off the Rawleigh and Watkins assembly lines. There were spices and flavorings to season foods, toiletries to soften and perfume the body, salves, liniments, laxatives, and cough syrups to relieve human aches and afflictions, and veterinary dips, disinfectants, and powders to treat the diseases of livestock. In 1921 the Rawleigh Company manufactured 125 different products; during a visit by the Rawleigh man customers could buy everything from blood-purifying tonics to vanilla flavoring and lemon extract, cinnamon and nutmeg, White Rose and Anna May

Bouquet perfumes, shampoo and face lotion, and a powder to rid chickens of lice.[32]

By 1912 both companies had established factories and branches in Memphis, Tennessee. Rawleigh boasted that its "beautiful and artistic" manufacturing plant there, surrounded by "a spreading lawn," was "the handsomest factory building in the South, being pronounced perfect by the State Factory Inspection." While the Memphis plant positioned the company to tap markets in the mid-South, a Richmond factory put the company within easy reach of retailers and customers in the upper South. Similarly, the Watkins company declared that its Memphis and Baltimore branches had "helped greatly in developing an enormous business in the southern states long before any other concerns operating by the wagon method had even made a start" and had "made the south solid for Watkins." Yet, the company announced it could still employ hundreds of salesmen because there was "a lot of virgin territory in the South awaiting the coming of 'The Watkins way.'"[33]

Both companies published manuals designed to entice men to join their sales teams and to instruct them in the most effective selling techniques. In 1906 W. T. Rawleigh wrote his first selling guide, and another followed in 1921, its 319 pages full of tips guaranteed to increase sales. Rawleigh's suggestions were exhaustive, and he anticipated every contingency. By 1926, the *Rawleigh Methods* guide filled 1,917 pages, and in bulk it resembled a thick brick. Meanwhile, Watkins published a similar—albeit slimmer—sales manual entitled *The Open Door to Success* about 1914.[34]

Like manufacturers of their era who "literally celebrated the 'mass' in mass production," Rawleigh and Watkins impressed employees and customers with the vastness of their companies. The Rawleigh enterprise, for example, began with expeditions to Madagascar, Sumatra, and other foreign lands in search of the most pungent spices, and then the company produced an extensive line of goods in large factories and coordinated complicated distribution systems that ran like well-oiled machines. The process ended when Rawleigh retailers established successful routes that delivered products to customers. Sketches of the Rawleigh and Watkins factories illustrated their publications, visually reinforcing the scale and scope of the operations and encouraging sales agents to consider themselves part of modern American businesses based on mass production and factory-made goods.[35]

Rawleigh and Watkins sought to distance themselves from unflattering images of itinerant merchants by stressing that their agents represented the vanguard of modern marketing methods. "Our men are not mere peddlers," insisted the Watkins Company. "They are high class salesmen with a real service to render the public." Rawleigh assured customers that the retailer of his products was "no stranger, or irresponsible, transient agent, or peddler to take your money and be gone," but "a neighbor and a friend. . . . He is not an outsider."[36]

Rawleigh coached his retailers in the principles of "scientific salesmanship," encouraging sellers to follow precise rules rather than relying on techniques and patter that they invented themselves. A good agent, according to the 1921 guide, "must be more than a mere order-taker—he must be a *creator and builder* of business, he must be able to influence and persuade others to buy his goods, he must create interest, then desire, and afterwards decision or action." With practice and experience, Rawleigh assured new agents, "you will be able to read character," design an effective presentation, and judge "your chances of making sales . . . almost instantly after you have met your customer."[37]

Rawleigh and Watkins manuals featured the stories of outstanding agents. Not surprisingly, their experiences endorsed company sales recommendations. Both firms encouraged retailers to make as many "Time and Trial" sales as possible. Customers who accepted goods on trial agreed to pay in full on the agent's next visit only if they had used the products and found them satisfactory. Agents shared tips on ways to dissolve customer resistance to buying and to collect money owed for products sold on credit. Indeed, testimonials usually took the form of long dialogues between the agent and his customers that read like scripts for novice sellers to follow. Although agents were encouraged to seek food and lodging with families along their routes, they were discouraged from indulging in evening entertainments; swapping stories or playing music was not part of the "modern" salesman's repertoire. Rather, the agent in these exemplary stories spent his time demonstrating products or preparing customers for sales that he would conclude before his departure in the morning.[38]

The Rawleigh Company counted both men and women among its customers, but testimonials by agents suggest that husbands were more skeptical of itinerants than were their wives. Men, according to these accounts, might refuse to buy from peddlers or "wagon men" because they preferred to patronize established storekeepers in their communities or suspected the quality of the products sold by itinerants. On the other hand, the trade literature portrayed women as customers who recognized the superiority of the Rawleigh spices, flavorings, and medicines.[39]

When a Rawleigh man or Watkins man pulled up to a southern farmhouse, he had a bag of sales tricks at his disposal—if he heeded the voluminous advice offered. The recollections of a Chatham County, North Carolina, woman whose grandparents bought from "the Rawleigh man" during the 1930s and 1940s suggest that agents followed some of the recommendations. Gladys Hackney Thomas recalled that the Rawleigh retailer who called on her family patiently described his products but did not dawdle. "He was always in a hurry," she remembered. "He took time to show his goods," but as soon as purchases were made, "he was on his way." The behavior of this Rawleigh man resembles the sales guide's directive to "be active and business-like" when making customer calls. "Do not loaf or loiter around one moment after finishing your business," the guide instructed. "Time is valuable, do not waste yours or compel others to waste theirs, but move on." In addition, the North Carolina agent met the Rawleigh dress code of "neat but not showy," a style with which rural customers would feel comfortable. "He didn't wear overalls like farmers did," Thomas remembered, "but he didn't wear a tie, either."[40]

Oral history accounts suggest that women especially welcomed the visits of Rawleigh and Watkins retailers because they provided breaks in household routines and opportunities to make autonomous purchases. Moreover, contrary to stereotype, women appear to have been discriminating customers who judged products and evaluated their efficacy with care. The Rawleigh man passed through Gladys Thomas's community about once every three months, and the women "were so thrilled" that they telephoned neighbors as soon as someone spotted him in the vicinity. He often arrived mid-morning, and Thomas's grandmother and aunts would interrupt their cooking to inspect the goods and make their purchases. The women treasured the Rawleigh salves and liniments that lined the family medicine chest, and Thomas's grandmother liked to add Rawleigh cake colorings to batter or frosting, a touch that set her cakes apart from the crowd on Sundays when a church dinner followed preaching.[41]

The Watkins agent who visited the eastern North Carolina home of Edythe Hollowell Jones in the 1920s became a source of amusement, but his products were highly valued. The family joked that the salesman planned his route in order to arrive at noon. "I've heard my mother say when he was in our neighborhood, she swore he heard our dinner bell ring and made it there in time to eat dinner," recalled Edythe Jones. "Oh, yeah! Mama'd always ask him to come in and have dinner and he'd eat." Irene Barber Hollowell selected "all kinds of spices," trusted Rawleigh salves to relieve congested chests and soothe burns, and "swore by" the veterinary tablets and powders she bought to cure her chickens of parasites. Examining the Rawleigh products brought to her door gave Irene Hollowell "an opportunity to make a few purchasing decisions" otherwise denied her when her husband filled the grocery list.[42]

Rural southerners used consumer goods to their own ends, incorporating purchased medicines into a pharmacopoeia that still included homemade remedies. New consumer possibilities did not eclipse older practices. Although Gladys Thomas's grandmother bought Rawleigh tonics and salves, she continued to gather wild mullein leaves and cherry tree bark to be boiled into a syrup for coughs. The fictional Dele Fuller purchased indigestion pills from the "Native Herb" man, but she and her female relatives searched the woods when they needed ingredients for a special salve that relieved aching muscles and joints or a tea that soothed sore throats. Even store-bought medicines were applied in dosages and according to methods that reflected the accumulated wisdom of farm women. Rather than being susceptible to bogus sales pitches, many farm women evaluated the curative powers of commercial remedies and made choices based on time-tested observations and experiences.[43]

Although Rawleigh advised salesmen to accept payment in goods instead of cash only as a last resort, agents nonetheless found themselves bartering with customers for whom eggs, chickens, or smoked meats were the only currency. In south Georgia, Clyde Purvis's mother swapped eggs for Rawleigh products. During the depression, Jessie Gosney's father ran a Rawleigh route near Stuttgart, Arkansas. Using hyperbole to describe the lengths to which her father went to cater to his poor clients, Gosney said that he "would get paid in dogs and cats and strawberries, rotten strawberries. Mother never will forgive him for that." Barter and mass marketing went hand in hand.[44]

Photographs suggest the popularity of Watkins products in the South and illustrate how farm families used company advertising as a kind of vernacular household decoration. In 1938, FSA photographer Russell Lee recorded a husband, wife, and six children in "a small, plainly furnished, but comfortable house" near New Iberia, Louisiana. The only adornment on the wall behind them was a Watkins calendar. When FSA photographer Jack Delano captured the image of "Mrs. Fanny Parrot, the wife of an ex-slave" near Siloam, Georgia, in May 1941, the subject posed before a fireplace. Above the mantel hung a Watkins calendar, and suspended below it was a Watkins almanac.[45]

As black southerners entered the market as consumers, albeit on unequal terms with whites, the Rawleigh Company took note. During a trip to his Memphis plant in the 1920s, W. T. Rawleigh became convinced that southern African Americans represented a lucrative market for the company's line of goods. When he published the 1926 edition of his sales guide, he entitled a chapter "Southern Collection Methods." Drawn largely from retailer testimonials, the chapter encouraged Rawleigh men to overcome reservations they might have about dealing with black customers and described the best ways to sell to and collect from them.[46]

The advice reflected a contradictory blend of entrepreneurial respect and racist suspicion and coercion typical of white southern country storeowners. "[T]reat the negro right," began the circuitous counsel of J. W. McCall of Tennessee. "Of course keep them in their place. Sell them the Products and let them know you expect their money and treat them as negroes should be treated. Treat them with the same consideration in a business transaction as white people." Although the sales manual cautioned southern Rawleigh retailers "against using old fashioned plantation managers' methods of extending credits and allowing Consumers to . . . pay only once a year" after fall harvest, agents who had profited from a large trade with black customers described taking out mortgages and notes as one way to ensure reimbursement for products sold on time. On some plantations, Rawleigh men consulted with managers and owners before selling to tenants and convinced them to agree "that the accounts . . . would be guaranteed by the manager and held out of [the tenant's] crop or wages."[47]

Rawleigh men who shared their methods of collecting payments revealed that they subjected black customers to bullying tactics never imagined for white buyers. A Georgia agent who tried to win the confidence of black clients and to treat them fairly nonetheless resorted to threats of suing "careless" blacks who failed to pay their bills. An Alabama salesman treated his recalcitrant patrons "firmly" and provided an example of a firm collection strategy. "One day," he recalled, "I had to get out of my wagon and whip one fellow to make him see my way, but from then on, I never had any trouble." To encourage retailers to tap the African American market, Rawleigh first contended that many black customers were as trustworthy as whites and then sanctioned legal and physical coercion should there be exceptions to this general principle.[48]

Rawleigh's philosophy toward black customers compounded the disadvantages that African Americans suffered in a marketplace where landlords and merchants commonly practiced usury and deceit. Yet Rawleigh did provide African Americans buying options that included the range of goods they could purchase and the site of consumption itself. While some retailers sought the consent and cooperation of plantation managers before approaching tenants, others met black buyers in places that blacks controlled and that were away from the watchful eyes of whites. S. B. Pearson of Alabama, for example, delivered products to and collected payments from black customers at a local church or schoolhouse.[49]

Regardless of the conditions of trade, African Americans may have welcomed the alternative to general stores that Rawleigh men and other sales agents offered. Merchants who "furnished" credit and supplies to farm families often scrutinized their purchases in order to limit the amount of debt accumulated through the year. For example, the southern Mississippi store-keeper who "furnished" N. J. Booth in the 1930s and 1940s restricted the purchases that the black farmer and his wife could make, although they owned their own land. The merchant, Booth recalled, refused to sell his wife groceries that "he figured we could do without. She might [make a list that included] something to make a cake or something like that. He took that off. Flour and lard, he would leave that on. But something extra, he always looked at the bill and he took that off." The merchant "had to okay the bill every time my wife or Brother Henry's wife [wanted to buy anything]—and he did the poor white the same way." Wayland Spivey, an African American sharecropper in eastern North Carolina in the 1940s, described a similar experience. The landlord "would give me a purchase order to go to the store and get food, and that was all I could get."[50] Given such restrictions, black farmers might welcome a Rawleigh man and a chance to buy flavorings and spices to beat into cake batter.

Not all itinerants undermined racial hierarchies, but peddlers, rolling-store owners, and agents of national companies at least provided white women and African Americans with options that were denied to them at country stores. Stores flourished and fostered a new consumer economy and culture in the South's cities and towns; at the same time, an older way of buying and selling persisted and assumed new forms as itinerants "linked the remote countryside with the marts of trade." Too often, historians of commerce in the New South have treated stories about mobile merchants and their customers as picturesque artifacts of an earlier era. But taken seriously, these stories reveal that humble consumers shaped the marketplace and enticed sellers to offer goods on their turf and on their terms. Careful observations of transactions conducted with itinerant merchants add new layers of meaning to the picture of consumerism in the rural South and demonstrate that women and African Americans were agents in the marketplace who stretched its boundaries rather than buckled to its constraints. As Myrtice Crews and her boy Harry knew every time the Jewish peddler drove into their yard, the southern country store was but one constellation in the region's universe of buying and selling.

CHAPTER TWO

Anything She Could Sell

LURLINE STOKES MURRAY remembered the exact moment her mother decided to produce for the market. Born in 1915 on a tobacco farm in Florence County, South Carolina, Lurline had no arch in her feet. Her father "finally scraped up enough money" to purchase special shoes with metal supports, but the metal quickly wore holes in her socks. "Mama decided then, honey, that she had to do something," Murray recalled. "Mama said she decided that she'd get anything she could sell, and that's when she started to taking what she had and turning it into money." Butterbeans and peas, eggs and butter, blackberries gathered from the woods—all could be sold in the town of Florence in the 1920s to buy socks for a little girl.

Lurline's mother, Julia Benton Stokes, already had a reputation among neighbors for making good butter and buttermilk, and for having a little extra to trade. Black neighbors, in particular, patronized her. They started "bringing their little bucket [for the buttermilk]," Murray recalled, "and in it would maybe be a half a dozen eggs or what you reckon? Octagon soap wrappers. They would come during lunchtime, when we were at the house for our dinner. They'd send the children." Julia Stokes, in turn, redeemed the soap wrappers at a Florence furniture store for dishes and "the nicest pots," some of which her daughter still used fifty years later.[1]

When she decided to sell more widely, Julia Stokes quickly discovered a ready market for all kinds of produce. A neighbor who sold vegetables at a curb market in Columbia, South Carolina, turned to her for help one week when he found himself short of turnip greens. "So Mr. Wallace Jones came out there one day," Murray explained, "and he asked Mama, he said, 'Miz Julia, how 'bout selling me some of them turnip greens? I got orders.' And them was the sorriest things. Mama said, 'You mean there's anything out there?' He said, 'Yeah. Just cut the greens off and let 'em grow back out and there'll be another cutting in a few weeks.' So one thing [led] to another."[2]

While historians have described in rich detail the cultivation and sale of the region's major cash crops—cotton, tobacco, rice, and sugar cane—they have largely ignored women's production for market and failed to appreciate its economic significance. This oversight is somewhat understandable. The major commodities developed elaborate and easily documented marketing systems that allow historians to trace average yields and price trends with precision. Annual harvests and trips to market in late summer and early fall represented a dramatic time of reckoning for farm families. In bad years, farmers limped home with little to show for their labor; in good years, they left tobacco warehouses and cotton gins with cash in their pockets.[3]

Women's trade, by contrast, appears more ephemeral, fluid, and elusive. Its value might even appear inconsequential compared to the proceeds of staple crops. Local markets and individual sellers and buyers often determined prices. Therefore, assigning an exact value to women's contributions to farm family economies is far more difficult than calculating the profits from cotton and tobacco. Women pursued a variety of marketing strategies. They traded with neighbors who came to their houses; they sold door to door to patrons arrayed along routes in town; they sold to itinerant buyers known as hucksters who made routine stops at their homes. These decentralized transactions, occurring in domestic spaces, were rarely recorded and therefore cannot be followed down a paper trail.

An aura of timelessness envelopes farm women's production for exchange and obscures its dynamic nature and changing context. Certainly, women like Julia Stokes could trace their ancestry as traders back through several generations, and those ancestors had counterparts all over rural America. Women's participation in barter economies is a standard feature of American economic lore. Yet using soap wrappers as currency signals an important change in rural economic life, reminding us that in the early twentieth century mass-manufactured goods increasingly replaced homemade and that consumers could cash in Octagon soap wrappers for items such as pots, pans, and dishes that could not be produced on the farm. The transaction's directness masks its complexity, as people chose how to combine store-bought items with home production. The exchange also suggests how race and class shaped women's economic options. Perhaps the African American women who lived near the Stokes family had stockpiled soap wrappers to trade because they earned money by taking in washing. Paying for buttermilk with wrappers that manufacturers used as premiums to build brand loyalty reveals the intricacy of the trading economy during the first decades of this century. In what appeared to be a simple transaction, family, local, and national economies were joined.[4]

Recent studies of antebellum yeoman farmers, the informal economies of slaves, and household production among rural freedpeople encourage us to take small-scale trading—and its economic and political implications—seriously. For the white yeoman farmers of the antebellum South Carolina low country, women's production for household use and exchange provided a critical margin of safety and surplus that secured their families' economic independence and underwrote their men folks' claims to political equality alongside planter elites. Ironically, the yeomen who cherished their independence so highly relied upon the labor of dependent women and children, just as planters relied on the labor of slaves.[5] Not far away, by the late eighteenth century, slaves on coastal Georgia rice plantations had developed elaborate informal economies that relied on foodstuffs produced or procured on their own time. Although both male and female bondpeople participated in trade, sometimes openly and sometimes surreptitiously, women dominated the vending on the streets of Savannah and in the town's public market. The slaves who created the informal economies of low country

Georgia were bent on "ensuring the welfare of the family."[6] The informal economy was no less important after slavery. Black farm families accumulated resources that lay beyond a landlord's reach by raising their own food and trading surpluses and by pooling wages. Little by little, families accumulated cash and livestock. In addition to a shrewd use of credit, families invested the profits of household production in land and in institutions vital to the well-being of politically embattled African American communities, including schools, churches, and charities. Whatever independence and autonomy freedpeople enjoyed rested on the household economy's foundation.[7]

Women were central to the household economy, and the household economy remained central to the southern farm economy even as it commercialized in the twentieth century. By failing to take women's trade seriously, historians become unwitting collaborators with a gender system that has valued the work of women less than the work of men. But when we reattach the household to the farm, we restore a key part of the calculus of production and consumption that figured into farm families' strategies for survival. Women's production for market both complemented cash-crop agriculture and shielded their farms and their families against the vagaries of the market for staple commodities. Through their trade, women provided a continuous and relatively stable source of income that enabled farms to endure. Under the best of circumstances, the income that women generated sustained the family and freed profits from crops for reinvestment in the farm enterprise. For tenants and sharecroppers, women's earnings reduced the family's indebtedness to creditors and landlords. And for the women themselves, taking what they had and turning it into money mitigated their economic dependence upon men.[8]

Women's trade evolved as the regional economy developed and in response to the needs of their families and to their own personal goals. Memoirs of rural life, farm periodicals, and extension service reports abound with stories about women's production for exchange and sale. Home demonstration agents who began to establish curb markets for southern farm women in the 1920s testified to the value of the trade. That trade became even more significant in the long agricultural depression that stretched from the end of World War I to the onset of World War II. For many families, women's earnings eased the pinch of poverty, put groceries on the table, met farm mortgage payments and tax bills, and kept younger children in school. Regardless of their source or destination, farm women's earnings clearly greased the wheels of rural commerce.[9]

EARLY-TWENTIETH-CENTURY southern farm women's production for market grew out of their production for home use. While the spread of cash crops in the late-nineteenth-century South had undermined diversified farming, rural families still tried to meet as many of their own dietary needs as possible. Provisioning the family larder required the labor of both men and women. Besides growing crops for market, men kept a few hogs for home butchering and raised corn to feed livestock and to have ground into meal. Their cane patches also provided molasses in the fall. Women planted gardens and preserved vegetables and fruits; they raised chickens for meat and eggs and kept a cow for milk and butter. Although the farm family economy revolved around crops raised for market that were considered primarily the responsibility of men, a family's well-being depended in large measure upon the thriftiness of women and the portions of the household economy that custom assigned to them.

A family's ability to achieve the goal of self-provisioning varied by class, race, and the power to determine the crop mix. In the early 1920s, rural sociologists surveyed a thousand North Carolina farm families in three counties representing the state's major geographical regions. The study found that both black and white landowners grew more of their own food than tenants and sharecroppers and that, in general, white families fared better than black. Farmers in eastern and Piedmont counties where cotton and tobacco held sway raised fewer "home supplies" than their counterparts in the mountains where cash-crop farming and tenancy were less common. For the total area surveyed, landowners raised 87 percent of their "living," while tenant families raised 76 percent; white farmers raised 88 percent of their living, while black families raised 64 percent.[10]

North Carolina playwright Bernice Kelly Harris captured the marketing ingenuity of farm women hard-pressed to make ends meet. Fellow Northampton County residents inspired Harris's characters in her 1937 play, *Open House*, which chronicled the eviction of a white sharecropper family in eastern North Carolina. Harris based her protagonist, Mrs. Jernigan, on a neighbor known as "The Butterbean Woman," who tramped hundreds of miles peddling butterbeans, berries, and garden produce. Like Harris's neighbor, Mrs. Jernigan scavenged and saved, picking up and reselling glass jars discarded by bootleggers, twisting broom straw off of creek banks and peddling homemade brooms. "She is never idle," Harris commented about the character she had created.[11] Regardless of class or race, few farm women were.

Women turned a variety of goods into commodities for trade, gathering and marketing what nature offered as well as what they produced. Poorer women picked and sold berries that grew wild in the woods. John Dillard's parents were white farm laborers near the south Georgia town of Tifton when he was born in 1920. His mother's resourcefulness at foraging for berries and selling vegetables warded off want during his childhood. "In the summertime," he recalled, "she'd get out and pick blackberries and huckleberries and haul them up here to town and sell 'em. Back then she would get five cent a quart for blackberries and ten cent a quart for huckleberries. . . . A lot of times she'd have a Number 2 washtub full of 'em. She'd just set 'em in the buggy and come up here around the edge of town and sell 'em. . . . You'd be surprised how many people'd buy that stuff." Dillard's mother also sold turnips to customers in town and to those who visited their home. Not far away in southwest Georgia, Bessie Jones grew up in a black family of farm renters. As a child, she and her mother braved snake-infested thickets to pick huckleberries to can and to sell. Jessie Franklin Felknor's parents were white farm owners in East Tennessee, and her mother bought blackberries for ten cents a gallon from poor neighbor women. "Daddy said that it was terrible when people came in on the back side of the place and stole your berries and then brought them to you to sell," Felknor recalled. "But mother always gave them ten cents a gallon for their berries."[12]

Women put their entrepreneurial skills to work selling a variety of goods, but poultry and dairy products formed the backbone of most farm women's trade. A flock of chickens in the yard and a good milk cow or two could provide a surplus for market as well as food for the family table. Because milk

"Noontime chores. A Negro woman and her daughter feeding chickens on a tenant farm." (Photo by Dorothea Lange; LC-USF 34-20187-E, Granville County, N.C., July 1939, Farm Security Administration/Office of War Information, Division of Prints and Photographs, Library of Congress, Washington, D.C.)

cows required pasture and hay, they were more often found on middling farms. But flocks of fifty or fewer chickens scratched and roosted everywhere. According to the rural sociologists' survey of North Carolina farmers in the 1920s, for example, more families raised poultry than kept cows. White owner-operators had more chickens and cows than any other group, suggesting that the women of these families produced more for the market as well as for use at home than their counterparts in either more affluent owner-landlord families or poorer tenant families.[13]

Raising chickens on a small scale required little capital, and their care could fit easily into a farm woman's round of duties. Eggs and live chickens served as a ready medium of exchange; in fact, some women kept poultry with an eye toward trade as much as an eye toward the dinner table. Zetta Barker Hamby's white family, which owned a fifty-five-acre farm in northwestern North Carolina in the early twentieth century, balanced household production for consumption and for exchange. "Every family tried to have chickens for eggs and to eat but mostly to sell," Hamby recalled. "It was reassuring to have eggs and chickens to barter at the local store for sudden, urgent needs." Young fryers that would fetch a good price were judged too valuable for their growers to eat; instead, when Hamby's family wanted to enjoy chicken they killed old hens that had quit laying and boiled the tough meat to tenderize it before dredging it in flour and frying it.[14]

The products of women's labor entered the channels of commerce in a number of ways, ranging from casual exchanges to formal markets. Modest transactions occurred so constantly and so routinely that they might be taken for granted by family and strangers alike. In eastern North Carolina, women traded eggs with men who traveled from house to house peddling fish caught in the area's rivers. At the home of Roy Taylor, son of white tobacco tenant farmers, the "fish man's" arrival on Saturdays signaled that his mother's cache of eggs would be raided. "Go see if your Mammy has any extra eggs," Taylor's father would instruct him. "If she's got three dozen or so eggs, the dollar added to the egg money will be enough to git the speckled trout." Nellie Stancil Langley's white eastern Carolina family also patronized a "fish man" who accepted eggs in trade for perch and spots as well as the soft drinks that he kept iced down among them. As a child growing up on a

North Carolina mountain farm, Virgie Foster went to the store "a many a time to get my daddy a nickel's worth of chewing tobacco." She began the errand by checking to "see if the hens laid," because eggs served as currency.[15]

Storekeepers in country and town alike accepted eggs and live chickens in trade. Nellie Langley carried several dozen eggs to the store where the family ran an account and paid "on our bill, so we could get us at least a piece of cheese to eat. It was Christmas if we could get a piece of cheese; we thought that was something." While John Dillard's mother took advantage of seasonal markets for berries and greens in south Georgia, she could trade eggs and chickens for groceries at Tifton stores year-round. In anticipation of using chickens to trade for supplies at the community store, Zetta Hamby's mother confined chickens to a coop and fed them generously for a few days so they would gain weight and fetch a good price.[16]

Although the women might not have known it, they formed the first link in a supply chain that connected home producers, country stores, and wholesale markets. Rural storekeepers served as intermediaries between scattered small producers and wholesale merchants located in towns and cities. An egg trade was at the heart of the general store that John Ward owned and operated in western North Carolina in the early twentieth century. The ledger that the Watauga County merchant kept in 1914 documents daily transactions with customers and demonstrates that "the economic life of the community surrounding the general store was one based on eggs more so than cash." Customers covered most of their bills with eggs, and occasionally corn, and paid the difference in cash. No doubt, such trade patterns were common throughout the South.[17]

Men's names usually headed the Ward store accounts, thus obscuring the centrality of women's commodities to the balance of trade. An exception was a transaction on January 2 by Maggie Ward, who received 35 cents credit for "1-5/12 doz eggs" and purchased sugar, oatmeal, cloth, thread, and candy for 56 cents, leaving a balance due the storekeeper of 21 cents. More typical was the January 5 transaction by Ben Hicks. His wife or mother probably had raised the chicks that weighed 26 pounds and earned a credit of $2.38. Hicks used the credit to pay for postage and toward the purchase of coffee and a knife, valued at $2.72.[18]

John Ward's invoice books reveal the important role that eggs played in his accounts with wholesale suppliers. One East Tennessee dealer in particular, J. J. McQueen, advertised "butter and eggs a specialty" at the top of his stationery. Although McQueen occasionally purchased green and dried animal hides from storekeepers, items that men would have processed, women produced most of the commodities that he bought from John Ward. And the poultry and dairy products proved to be lucrative investments. On May 25, 1911, Ward sold McQueen 3 bushels of beans, 27 pounds of butter, 210 dozen eggs, and 300 pounds of chicken for a total of $61.67, nearly $10 more than the week's order of coffee, flour, fertilizer, and boards from McQueen. A month later Ward sold $30.39 worth of eggs, butter, hens, chicks, and "cox," just $3 shy of covering the cost of the brown and granulated sugar, rice, tobacco, salt, and flour that he ordered from McQueen that week.[19]

The quantity and price of eggs rose and fell during the year, with shorter winter supplies commanding premium prices nearly double those of the spring. The thirty dozen eggs that Ward sold to McQueen for 28 cents each in late November of 1911 helped stock his shelves with peanut butter dandies, coconut bon bons, crystallized jelly squares, chocolate, and jelly beans in time for the winter holidays.[20] Women's poultry flocks put candy in children's Christmas stockings.

Throughout the year, McQueen kept the Watauga Falls storekeeper posted about fluctuations in poultry and egg prices. "It is pretty hard to tell what the price on eggs will be," McQueen noted at the bottom of his February 9, 1911, invoice, "but not far from the price paid you. about 14c Per. Doz. loose 15c cased chicks are pretty strong and if I were you would not miss anny [sic] of them. I am quoting 11c on them but can afford to raise this price some if neccessary [sic]." Ward stockpiled eggs during the next month, and by late March had 250 dozen for sale. McQueen advised Ward that Bristol, Tennessee, poultry dealers had forecast lower prices the following week. "I will protect you on what you have on hand," McQueen assured him, ". . . but I would not think it advisable for you to buy expecting more than 11c loose 12c cased for next Wk. [Of course,] they may not change the price on them but it looks now like they will." In a handwritten postscript the wholesaler advised Ward that the 250 dozen eggs would bring 13 cents if cased. The products of women's labor were the lifeblood of John Ward's store.[21]

Nestled on the border between North Carolina and Virginia, the country store where Zetta Hamby's family traded was the first stop on a long trip between the producers of poultry and dairy products and their consumers. After weighing and buying live chickens, the Grassy Creek storeowner penned them in a coop until representatives of a wholesale house collected the chickens and conveyed them to a railroad station from which they headed to Virginia cities and beyond. The butter that Grassy Creek women churned and traded at the store—where it was held in wood barrels—might wind up as far away as Baltimore. A neighbor of Hamby's who accompanied a commission merchant to the mid-Atlantic city reported that to test the quality "the wholesaler ran a rod down into a barrel of butter all the way to the bottom, took it out and tasted it before buying it." Hamby speculated that the Baltimore dealers sold the Grassy Creek butter, not directly to consumers, but rather to a creamery where it was reworked and packaged for retail sale, eventually appearing on the city's dining room tables.[22]

When North Carolina merchant O. S. Edmonds took the "rolling" version of his store into the hollers of Madison County, women used eggs, live chickens, and butter to pay for the groceries they bought. Clerks carried crates to hold the dozens of eggs they accumulated on any given day. When egg production peaked from February through May, some women presented as many as thirty dozen eggs to trade in a week. Should the women not need as many groceries as their eggs were worth, Edmonds paid the difference in cash. A wholesale dealer in Asheville pegged the price of eggs, and Edmonds usually paid the women about five cents a dozen less than he stood to receive.[23]

Hucksters traveled the South, paying cash for poultry and dairy products to the farm women who greeted them in their backyards. In northeastern North Carolina, G. Emory Rountree earned much of his living between 1932 and 1960 working as a huckster. The Gates County native traveled the countryside buying eggs, live chickens, and rabbits, which he then sold in the Tidewater Virginia area some forty miles away. Rountree's older brother had preceded him in the huckstering business, picking up eggs and chickens in a mule-drawn cart. In 1932 Rountree started using an automobile to cover the territory he claimed, from his home in Sunbury to the nearby communities of Hobbsville and Trotville, and he later graduated to a truck. Although other buyers established routes in the vicinity, Rountree considered himself "one of the big hucksters" because he drove a larger-capacity truck, covered a wider area, and picked up eggs and chickens several days a week.[24]

To launch his trade, Rountree "started out with people that I knew maybe, and then they would tell somebody else. . . . I was one of the few that would get as many as a hundred crates of eggs a week." Although Rountree also bought old hens, young fryers, rabbits, and pork products, eggs were the mainstay of his business. In addition to buying from individual customers, Rountree purchased eggs that storekeepers accumulated when customers brought them in for swapping.[25]

The eggs and chickens that farm women in Gates County produced wound up for sale at a variety of outlets. Rountree sold some to retail customers at the city market in Portsmouth, Virginia; he sold some directly to independent grocers in the Tidewater area; and he sold to wholesale dealers in Norfolk. To take advantage of the highest prices, Rountree housed crates of eggs at rented cold storage facilities in Norfolk for several months at a time.[26]

Men like Edmonds assumed that women were unaware of prevailing market prices. "We usually knowed a day or two ahead of time" the price eggs were bringing, the merchant recalled. "We went to town. But back up in these hollers, they didn't know what the price of nothing was. They just mostly took your word." But there is evidence that at least some women knew the value of their products and guarded against the possibility of a huckster's duplicity. One of Rountree's customers, for example, insisted on weighing her chickens on her own set of scales, which she judged more reliable than his.[27]

Women also chose to control their own retail operations, thus sidestepping intermediaries. Some sold occasionally, while others developed regular routes of customers. Born in 1909, Sarah Webb Rice grew up in a black sharecropper family in Alabama, where they raised livestock and kept a garden. During lean times "Mama would peddle eggs and peas" and blackberries in the town of Batesville, Rice recalled. "We would sell those good old shelled peas for fifteen cents a gallon and were glad to sell them for that." After her mother died, Margaret Christine Nelson lived with her grandmother, a former slave who farmed in Clarendon County, South Carolina. As a child in the 1920s, Margaret drove a horse-drawn wagon to Summerton and sold her grandmother's butter to white patrons for twenty-five cents a pound and butterbeans and peas for ten cents a quart. "I had my special customers," she recalled. C. B. Player Jr. grew up in a white landowning family that was far more prosperous than most of their Clarendon County neighbors; hired laborers milked the cows for his mother, but each week she processed, molded, and delivered her prized butter to loyal buyers who lived in town.[28]

Julia Benton Stokes and Lurline Stokes Murray developed a mother-daughter partnership that spanned three decades and catered to customers in Florence, South Carolina. Beginning in 1925, at the age of ten, Lurline accompanied her mother into town twice a week to dispose of buttermilk, eggs, and seasonal fruits and vegetables. These sales remained a vital part of the farm's income even after Lurline and her husband, J. W. Murray, bought a nearby 390-acre farm in 1938. Lurline and a tenant tended the land while J. W. worked as a machinist for the Southern Railway. By the 1940s, Lurline and her mother supplied twenty-five regular clients and restaurants with eggs and dressed chickens. Customers reserved standing orders, and if their needs changed they would advise Murray to adjust the quantity to be delivered the next week. Stokes and Murray sold to a wide spectrum of patrons. Their clients found them by word-of-mouth recommendations and ranged from doctors, lawyers, bank presidents, and preachers to J. W. Murray's blue-collar coworkers at the railroad roundhouse. "I have never asked nobody to buy something," declared Lurline Murray. "If you've got something and people know what you've got, it'll sell itself. I've sold to some of the richest and the poorest."[29]

Farm women used a number of outlets to market their goods. During the early twentieth century, some took advantage of the burgeoning mail-order business to reach customers more than a day's ride away. In 1915 a group of rural women in eastern North Carolina "combine[d] business and pleasure" by collecting and shipping their eggs in bulk and developed a market for homemade butter by advertising in newspapers and using parcel post to sell directly to city housewives. Properly packaged and chilled, all manner of products could survive a day's journey through the mails. In 1916 a Tennessee farm wife, Mrs. W. H. Alexander, used personal connections and parcel post to enter the distant Birmingham, Alabama, market for butter. Before developing her city trade, Alexander had sold small quantities of butter to neighbors and peddlers "who came around each week never paying over 15 cents a pound." Dissatisfied with "such prices," Alexander convinced her husband to buy a Jersey heifer and studied how to churn, mold, and sell superior butter. A friend who lived in Birmingham engaged customers for her, and within two months she could hardly meet the demand for butter that brought thirty-five cents a pound. Alexander expanded her dairy herd to four cows and eventually mailed twenty-five pounds of butter every Tuesday morning, each pound neatly sealed in a box she made herself and secured with twine. Although her customers could have bought cheaper butter, one patron told her, "'it is not fixed up nicely like yours.'" Alexander concluded, "'Fixing up things' means something."[30]

Women's financial contributions to their families were not lost on southern advocates of rural economic development. Beginning in the 1910s, agrarian progressives who envisioned more prosperous farms and homes sought to harness women's productive enterprises and developed institutional outlets for their goods. Assisted—or in some cases thwarted—by state home demonstration agents, farm women encountered government regulations for the first time. Even as some rural women appreciated agents linking rural producers to town consumers, they also found their local markets under scrutiny as home agents set new standards for the cleanliness and appearance of goods sold. In 1922, for example, the home agent in Transylvania County, North Carolina, reported that a local college refused to buy butter from anyone until she had "visited the home, approved the sanitary conditions and shown them how to pack butter."[31] Marketing proved a contested arena where rural women and rural reformers had to reconcile different agendas, assumptions, and goals.

Home demonstration club curb market, Buncombe County, North Carolina, 1931.
(28919-C, Record Group 16-G, U.S. Department of Agriculture, Box 235, Markets-Roadside folder, National Archives and Records Administration, College Park, Md.)

During the 1920s, home demonstration agents actively organized curb markets in county seats across the South. Responding to the downturn in the agricultural economy, women's markets expanded at the same time that prices for cotton and tobacco began to decline. In the late 1920s and early 1930s the regional network of home demonstration markets grew. Women in Alabama, South Carolina, Virginia, and North Carolina praised these markets for helping them support their families as crop prices went into a tailspin. A handful of North Carolina markets in 1922 mushroomed into thirty-six markets by the mid-1930s, and other southern states followed a similar pattern. In 1932, North Carolina women who belonged to home demonstration clubs sold items valued at some $184,000; the next year earnings climbed to $274,000. In 1936, some 1,400 weekly and seasonal Tar Heel market sellers grossed nearly $263,000, and two years later 1,700 women sold $309,150 worth of goods at organized markets. All the while, clubwomen reported that they earned thousands more dollars from sales to individuals, schools, and hotels.[32]

A market paired the farm skills of rural women with the civic know-how of home demonstration agents. Women who wanted to sell literally at the street curb formed a committee of home demonstration club leaders to visit the "city fathers" in the county seat to secure parking spaces on a busy thoroughfare for sellers one or two mornings a week. In some counties, women displayed their goods in the basement of the county courthouse or used a tobacco warehouse in the off-season. Once a market gained popularity, sellers might seek a more permanent space. In Mecklenburg County, North Carolina, for example, the courthouse in Charlotte served as the first market headquarters in the early 1920s. Home demonstration clubwomen soon appointed a committee to lobby county commissioners and city council members for an appropriation for "an adequate means of bringing them in direct contact with the consuming women of the city," and the officials responded with a $30,000 donation for a market building.[33]

If farm women needed markets to supplement their incomes, home demonstration agents viewed the markets as tools of rural reform and an opportunity to uplift farm women. Agents believed that by bringing farm women into closer contact with more "refined" town women, curb markets would inspire sellers to improve their manners and personal appearance. Like many of her colleagues, the home agent in one eastern North Carolina county worried that the "narrowly circumscribed horizons" of country life fostered "sluggish, apathetic" families and offered "no models from whom to carry away hints of poise in speech and action." Mingling with town buyers at the curb market was one remedy for this presumed problem.[34]

Women who sold at the home demonstration curb markets had to abide by regulations that determined all aspects of sales, from the prices they charged to the appearance of their assigned tables and stalls. In this regard, they joined a long tradition of public markets the world over that established rules to govern vendors and to protect consumers. Before the market opened, the home agent or her representative canvassed local merchants to

"Customer buying at stall of club market, Wayne County, North Carolina. Goldsboro, North Carolina, July 1931." (s-15102-c, Record Group 16-g, U.S. Department of Agriculture, Box 233, Markets-Farmers'-1 folder, National Archives and Records Administration, College Park, Md.)

determine "cash and carry" prices and then set curb-market prices accordingly. Selling could begin only after an opening bell rang and had to stop when a closing bell sounded. While customers surveyed the displays of dressed poultry, eggs, cakes, vegetables, and fruits, sellers had to remain in their designated places and to keep those spaces tidy. When the market ended, sellers submitted a record of their earnings and contributed a share of their proceeds to cover market maintenance and promotion.[35]

Some marketing committees went so far as to limit competition and to modify behavior. The first rule of the market in Carteret County, North Carolina, was simple: "No peddling." Women could not make deliveries to private homes or boarding houses on days that they sold at the curb market, nor could they sell to grocery stores until the market had closed. To take advantage of one sales outlet, the women had to forfeit others. In addition, Carteret vendors had to wear washable white dresses whose uniform appearance suggested cleanliness and order to buyers. The home agent coached women to talk "in a mild and pleasant tone of voice" when approaching potential customers. Market rules in Beaufort County, North Carolina, forbade sellers to "call to customers or in any way try to induce them to buy from them. After they stop at your *table offer* them the things you think they will like."[36]

Home demonstration markets practiced racial segregation. The bylaws that governed the market in at least one North Carolina county stated explicitly that only whites could sell under its auspices, and other markets followed the color line. Nonetheless, in the 1920s and 1930s, black home agents in North Carolina acted as intermediaries between producers and buyers, encouraged clubwomen to participate in municipal markets open to sellers of both races, and occasionally organized small curb markets of their own.[37] When black clubwomen in Columbus County decided in 1930 to coordinate their sales, agent Sarah J. Williams gathered all they had to offer, parked on the street in the county seat of Whiteville, and turned her own car into a curb market. In 1932, a club member in Alamance County worried because "she could not call on her husband for the family necessities" after crops failed the previous year; in response, the agent advised her to sell her garden and poultry products in a nearby town. Combined with cash earned for prize-winning exhibits entered at a county fair, the club member reported adding $128 to the family budget.[38]

Although most black club members in North Carolina "sold by house-to-house canvassing to regular customers," women in two Piedmont counties established markets. Denied the government resources that subsidized white women's sales, in 1937 black clubwomen in Rowan County took matters into their own hands and built a market shed beside a major highway. Two years later, club members mounted a campaign to improve the building's appearance and offered produce that they had cleaned and sorted by type and size. The women divided the $76.25 that they realized from May and June sales

"Roadside market furnished and operated by Negro club members, Charleston County, South Carolina, 1932." (Photo by G. W. Ackerman; S-15869-C, Record Group 16-G, U.S. Department of Agriculture, Box 235, Markets-Roadside folder, National Archives and Records Administration, College Park, Md.)

according to the amount of produce each contributed. Finally, they pooled some of their profits so they could buy a used icebox for $10, which allowed them to increase sales of butter, milk, and other perishables.[39]

To attract customers to the markets, some home demonstration agents used modern promotional tactics. They advertised in newspapers, distributed circulars that described goods for sale, and solicited endorsements from local doctors, businessmen, and teachers. Most markets held a prize drawing at mid-morning as a way to encourage customers to stay longer and spend more money. And vendors learned that they themselves were part of the promotional package. Personality, self-presentation, and "salesmanship" mattered nearly as much as the quality of the food for sale. "Wear a smile," the Beaufort County, North Carolina, agent advised market women, and "have a pleasant word for your customers. Thank them or ask them to come back. Let this be natural and not over done." In 1931, the agent in Carteret County, North Carolina, praised women who had "learned very quickly to gauge the customer to a nicety and be neither too insistent nor too listless in addressing her." The change in one woman's approach especially impressed the agent: The first day that she sold at the market, she pleaded with a visitor, "I haven't sold anything yet; don't you want to help me out?" The woman soon changed her pitch, and could be heard to say, "Good morning, Mrs. Thompson, can't I help you plan your dinner today?" Her sales soared as a result.[40]

The income-earning strategies that Tom Cunningham's mother pursued combined sales at home demonstration curb markets in South Carolina with sales to individual customers. Tom's mother sold first at the curb market in Greenville and later in Darlington after the family moved in 1929 in search of better land. In fact, Cunningham speculated that the home demonstration market provided the family some continuity and a place to meet people in a new community after their relocation from the upcountry to the coastal plain. "I'm sure it must have been a big transition for my parents and my older sisters and brothers to just pick up and leave and move down here where they didn't know anybody and set up a new operation," Cunningham explained. "She got right into the swing of things."

At the market, Cunningham's mother sold milk, butter, and eggs year-round and offered other goods for sale as the seasons changed. In the spring, she supplied tender greens and tangy onions. In the summer, watermelons, cantaloupes, and garden vegetables filled her market table. In the fall and winter, she sold cured hams, sausage, and peanuts. Cunningham's mother also maintained a route of customers in Darlington. The family had a telephone, and townsfolk called the farm about three miles away to put in their orders. On their morning trips to school, Tom and his siblings delivered milk, eggs, and butter to town doorsteps. "My mother," Cunningham recalled, "knew exactly who got milk what day and how much."[41]

Farm women in North Carolina used home demonstration curb markets to their own ends. Although agents frowned upon women who peddled, they eagerly encouraged women to invite their private customers to patronize the curb market. In turn, women who sold at a curb market on Saturdays made contacts there that became private customers on other days. An Anson County woman, for example, gained a following at the market for her cakes and beaten biscuits. Soon she was catering parties and weddings and providing a local drug store with dozens of small cakes each week. Most clearly, if curb markets failed to meet their needs, women abandoned them. In Cleveland County, sellers decided they could make more money through deliveries to individual customers, and the curb market died as a result. In Cabarrus County, the market withered in 1933 when sellers squabbled over rules and regulations and refused "to stop their house to house canvassing."[42]

Given the available sales options, women weighed their comparative advantages. Trading with storekeepers offered benefits and drawbacks. Because men often took the products of women's labor to stationary stores, women had less control over the transactions. The contributions that wives and daughters made to the household economy disappeared as they were claimed under male names on ledger pages. Moreover, storekeepers extended credit or trade in merchandise rather than paying cash, guaranteeing themselves customers for dry goods and groceries as well as a stream of produce that could be sold for lucrative prices. Nonetheless, storekeepers accepted small quantities in trade at any time, and by some accounts, were not too particular about the quality of the eggs they collected. Hucksters paid cash, were more prone than storekeepers to recognize and reward differences in quality, and saved the producer a trip to market. Women who lived close to towns where sufficient demand supported regular routes of customers, who could maintain steady supplies, and who had time to deliver their commodities might receive higher prices.[43]

Many women, whether white or black, enjoyed marketing in a lively public venue where they mingled with friends and strangers, managed the transactions, and turned their private work into social labor. Sellers engaged in good-natured rivalries, and women who baked mouth-watering cakes, preserved clear jellies and crisp pickles, or turned out well-molded butter reveled in the reputations that their products and their skills earned them. And many women believed they made more money at the curb market than anywhere else. Mrs. Foster Ricks, who sold at the Henderson County, North Carolina, market in the late 1930s, assessed the merits of various selling strategies. Peddling house to house "wasn't a pleasant task," Mrs. Ricks told an interviewer for the Federal Writers Project, and sharp-trading boardinghouse keepers "would try to git our stuff for almost nothin'." Miserly grocers, she complained, "only paid us about one-third o' the market price an' even then we'd have to take it out in trade at their stores." But at the home demonstration market, the women received "regular prices for everything." In addition, farm wives and mothers who could boast market earnings might influence how the family allocated its labor, and husbands and children who helped them set up and sell also witnessed the value of their labors firsthand.[44]

The proceeds from women's farm trade often made a substantial difference in a family's budget. Mothers bought school clothes for their children and put extra food on the table with their butter and egg business. Walter Anderson's parents were white farm owners in East Tennessee when he was growing up in the 1910s; his mother sold eggs, chickens, and turkeys to hucksters and used the proceeds to buy food and groceries that the family did not raise. Anderson recognized that his mother would "get as much money" for poultry products "as we would for the corn" grown on the mountain farm. He considered her earnings "a great help to us. If it hadn't been for her, we couldn't've hardly survived, I don't guess. It would've been rough, six children, you know, on a little forty-acre farm. It would've been pretty rough."[45]

Even small sums of cash stretched a long way in the 1920s and 1930s. A half-century later, Lurline Murray could remember precisely how she spent the dollar that she earned as a teenager by picking twenty quarts of blackberries for the wife of a Florence, South Carolina, bank president. "You could buy rice, three pounds for a dime," she began. "You could buy sugar, five pounds for nineteen cents. Salmon was two cans for a quarter back in them days, if you could get your hand on any money. You could get a pound of coffee for fifteen cents, two pounds for a quarter. I bought a little box of macaroni. I bought a little piece of cheese. When I got it added up, it was eighty-nine cents. I had my tithe and one cent over."[46]

The fact that poultry products were available nearly year-round rather than seasonally enhanced their worth. In north Georgia, Aubrey Benton's

"Farm women who are members of the Wilson market arriving with the produce which they have for sale, Wilson County, North Carolina, May 1940." (s-6354, Record Group 16-G, U.S. Department of Agriculture, Box 233, Markets-Farmers'-1 folder, National Archives and Records Administration, College Park, Md.)

mother kept "twenty-five old hens and a rooster or two" that provided eggs for trade. Eggs were most plentiful in the spring, when cash reserves had dwindled and farmers needed to finance crops. "Then you only had a little paycheck," Benton explained, "and that was in the fall of the year after you gathered your cotton. A lot of . . . the banks, they hesitated to loan you anything over fifty dollars to make a crop on. So them eggs was the main point in the spring, swapping for [groceries]." In South Carolina, A. M. Harrington's parents were white sharecroppers who raised tobacco and cotton in the 1920s. His father took eggs produced by his mother's chicken flock to town "about once a week" and used the money to buy groceries and clothes "for the kids to go to school." As Harrington observed, "You could almost grow a crop and wouldn't hardly see any money 'til you sold a crop." But a farm family could count on poultry products nearly year-round.[47]

Women stayed attuned to the market, calculating when eggs would be sold and when they would be eaten and negotiating a tension between self-provisioning and earning cash or credit. When income earning took priority, sales of goods deprived some families of food, or left them food of inferior quality to eat. Fredda Davis, raised on a mountain farm in North Carolina, remembered, "of course, we'd eat eggs once in a while, but we'd save them to go take to the store, to sell those eggs to buy these groceries that we had to buy, see." For Ruby Byers, who grew up in a white tenant farm family in north Georgia during the 1920s, seasonal supplies and prices often determined whether she ate eggs or traded them. During the winter, when the fewer eggs that hens laid brought higher prices, "you didn't eat the eggs for breakfast. You know, they wasn't that plentiful. Eggs was a treat when you had eggs to eat. And there was times when hens would lay more eggs, and prices were real cheap, and we'd eat the eggs. I mean, they'd take enough to get the soap, and coffee, and sugar, and stuff like that. . . . But the eggs was a precious commodity because as long as the old hen laid you could kinda count on falling back on that to get some things." Virgie Foster grew up in a family of eleven children in Wilkes County, North Carolina; her husband's family was much smaller. Whereas Jim Foster's mother cooked fried chicken "pretty often," Virgie's mother usually prepared chicken and dumplings because that dish made the meat go farther. Her husband's family, Virgie Foster explained, "didn't take [their chickens] to the store like we had to."[48]

During the bleakest years of the Great Depression, the income that farm women generated made a crucial difference for many families. In hindsight, Tom Cunningham could not estimate what proportion of the family economy his mother's earnings represented because he was a child during the heyday of her sales to private and curb-market customers in Darlington. He did know, however, that his mother "helped to keep body and soul together back during those depression years, when there just wasn't any money. I know that she worked at it diligently and that my father worked at the farm and between the two of them, they kept the bills paid and we were fed well; we had adequate clothing even though sometimes we had patches on our elbows and knees. But we fared well. We got along real good. And I know it took both of them to do it."[49]

The Cunningham family was not alone. White and African American marketing women in North Carolina testified to the importance of their earnings. In 1926, a Beaufort County woman turned to the curb market after a hailstorm shredded crops. The money she earned from selling broilers, eggs, butter, and vegetables covered the family's "living expenses" and reserved any profits from the next year's crops for reinvestment in the farm. In Buncombe County the white home agent concluded in 1930 that many of the women were "making the incomes of the family." A woman who sold at the Alamance County market said that the $500 she earned in 1932 "has helped us to keep Old Man Depression away." Similarly, the Durham County agent in 1933 observed that the curb market provided "the only cash income of some of the families and has paid the taxes and saved many farms." The same was true in Carteret County. "There is no way of knowing," the agent remarked, "just how much the market aids in paying taxes and buying school supplies." A seller in Richmond County told the agent that the market "has meant everything to me and my family. . . . I have bought all the food not raised on the farm for our family of seven, school books and supplies for the children in school, and all the clothes for the family. . . . In this way, I am helping to make money and take care of the family and it leaves all that we make otherwise to be applied to the debt on our farm." Another Richmond County woman saved a dollar a week for each of her eight children "to help with their education." In Robeson County, after the black home agent encouraged a club member to record her milk and butter sales, the woman tallied up a take of $132. Once she realized her contribution to the family, the woman told her husband that "she was through signing away her rights," and refused to endorse a mortgage on their home in exchange for credit. Instead, the couple used her earnings to pay cash for fertilizer.[50]

How much women's marketing mattered is suggested by figures that federal census takers gathered in 1929. While the earnings from dairy and poultry products never approached that from cereals and all other field crops, neither were they small change. All told, when South Carolina farm women like Tom Cunningham's mother sold their butter, cream, milk, chickens, and eggs, they earned some $12.3 million; meanwhile, the value of grains, cotton, and tobacco amounted to about $115.5 million. Across the border in North Carolina, farm families realized some $20.8 million from poultry and dairy sales and $198.5 million from cereals and field crops.[51]

Statistics hint at women's contributions to farm incomes; stories interpret the numbers in touchingly human terms. In oral history narratives, older southerners often remember the ways in which women's earnings were spent with a loving precision unmatched in stories about men's earnings. The income that women generated sometimes provided the few small luxuries that rural children enjoyed. In the early 1920s when it came time for Jessie Felknor to graduate from high school in East Tennessee, her mother's chickens paid for her class ring. "I asked Daddy for the money," she recalled. "He said, 'Sis, I don't have any.' He said, 'I told you that I didn't think we'd have enough money to do it.' So mother sold eight old hens; no, she sold sixteen. We'd catch 'em and tie strings around their feet and we'd tie 'em in bunches. And we'd tie four of those hens together. I know we took enough hens to the store that I paid for my class ring out of her hens 'cause Daddy didn't have any money." In eastern North Carolina, Roy Taylor paid his way into movies and drank fountain Cokes in Wilson during the 1930s with money that his mother gave him from her egg earnings. In the early 1930s, Ruby Byers's mother gave her daughter eggs to swap for the fixings for sandwiches that she and her siblings ate on an end-of-school field trip. "In order for Mama to have something for us to take on that picnic," Byers remembered, "she had to gather up the eggs and send us to the store to get something to make us a sandwich out of. They paid you, say, fifteen cents a dozen, and this particular time we got crackers and peanut butter, and my aunt made us some cupcakes and put chocolate on 'em and, boy, I mean we had a treat!" In southwestern Louisiana, Walter and Hattie Young's first child was born in the early 1930s. Strapped for cash, Hattie Young sold eggs to buy a pretty pair of baby shoes trimmed with tiny bows that the couple had admired at a local store. "We paid a dollar and forty cents," Walter Young recalled. "We sold fourteen dozen eggs to pay for that at ten cents a dozen." After the baby outgrew them, the shoes served as hand-me-downs for the next child. Eventually, the Youngs could afford to have the shoes bronzed and still displayed them proudly more than half a century later.[52]

Earnings also mattered to women who reported that their income had given them a say in farm decision making and family finances that they had never enjoyed before. When prices for the tobacco that her family raised in Robeson County, North Carolina, began to fall, Mrs. R. D. Graham decided she had to do "something to help my husband make a living." How could she earn money and remain at home with her children, too? The couple's two cows and calves provided the answer. Mrs. Graham began selling butter, cream, and buttermilk in the town of Lumberton on Saturdays. When her sales paid off, she convinced her husband to invest in better livestock, and by 1936 the couple owned five good cows. "My husband does the milking and helps with the churning when he has the time," Mrs. Graham reported, "and I do the rest. Marketing and all." After ticking off how she spent her earnings—to buy groceries, gasoline, clothing, and dairy feed not raised on the farm, to pay Grange membership dues and insurance premiums, and to drop a donation into the church collection plate—Mrs. Graham counted the greatest payoff to be not having to ask her husband "for money every time I needed something." A seller in Carteret County, North Carolina, confided to the home demonstration agent, "This is the first time since I married, twenty-one years ago, that I haven't had to call on my husband for every nickel I spend; and you don't know how good I feel to know I have a little money of my own."[53]

Agents noted a subtle but perceptible shift in gender relations. "Many women who beforehand had no occasion to learn are now well versed in handling the family finances," the Carteret agent reported. "Indeed, before the market opened only four market [women] had bank accounts. Now three-fourths of them are writing their own checks." A Fayetteville merchant told the agent in Cumberland County, North Carolina: "That market you have over there must be a good thing. The other day a man came in here wanting a plow point, but did not have the cash for it. He went across the street to the basement of the courthouse, where the woman's market was in free swing, to interview his wife, and soon came back with the money for his plow point." By 1933 Rosalind Redfearn, the agent in Anson County, North Carolina, declared, "When I first went into home agent work, the little cash that farm women found it possible to earn was always spoken of as 'pin money.' But with ever increasing profits from ever widening sources, we have now reached the point where we no longer say 'pin money.' Instead we say, 'the farm woman's income.'"[54]

Although practices varied, many women controlled the money they earned. Mary Wiggins Turner's father owned farms and ran a store in northeastern North Carolina during the early twentieth century, and her mother raised chickens. "Chickens and eggs was all the money the women had," Turner recalled. "Now, they weren't allowed to write checks or have any money. My father was the boss, and he was the man that did the check writing." But her mother "kept all the money she had from the chickens and the eggs and things like that. She spent it in the grocery bills or anything that she wanted." In Walter Anderson's family, his mother was "the boss" of her earnings from poultry sales. The children often benefited from her largesse. "Whatever we wanted, she got it. She was awful good-hearted, and she'd not let none of us go in want if she could be any help in any way." Although the profits that C. B. Player's mother earned from selling butter "was her money," he conceded that she used the money "for buying us anything she wanted to." In some families, however, the collective pot was too small to assign to an individual. Although the eggs from Ruby Byers's mother's chicken flock secured groceries for the tenant family, "there wasn't no 'my money and your money.' It was just lucky if 'we' had money."[55]

In East Tennessee, the income that Della Sarten earned from selling dairy and poultry products guaranteed her a measure of autonomy after she married in 1927. Will and Della Sarten grew grain and raised cattle, and she separated cream and churned butter from the milk that half a dozen cows produced and kept plenty of good laying hens. Rolling stores, commercial creameries, town merchants, and neighbors bought from her. "That's exactly how I made my living," Sarten declared. "I'm not a stretcher, but I bet I've sold two thousand pounds of butter." She used the money to purchase groceries, children's clothes, and household amenities and to contribute to the general farm operation. Accustomed to having money of her own before marriage, Sarten controlled a separate bank account all of her adult life. "I never did have to beg [my husband] for no money," Sarten said. "I'm not a-bragging, but I never did have to." Sometimes, however, the shoe was on the other foot, and Will Sarten borrowed money from his wife. "Yeah, lots of times he'd run out of money, come in and say, 'I've got to buy some wheat at the store. You've got any money you'll let me have?' I'd just hand it over to him. It was his as much as it was ours. But that's one thing, . . . I know a lot of husbands wouldn't even let their wives have a dime. Well, you know the difference in people. That's one good thing that I've had a say-so over my milk and butter and eggs ever since I kept house. I know a lot of men, just as soon as [their wives] sold it, they got the money and did what they wanted to with it, and the women just had to go by. But I've been pretty lucky. I can say that much."[56]

The proceeds from women's trade could guard against profligate husbands. Early in her selling career, Julia Stokes deposited a portion of the money that she earned into a savings account at the urging of her daughter Lurline, who feared that her father would leave them poor and dependent. "My daddy drank," Lurline Murray confided. "And every time he thought Mama had a little bit of money, he'd make her mad. You know how he made her mad? Tell her she give it to her people. And she'd get mad and say, 'Here. Take it.' It went on thataway and went on thataway. [That was about 1930.] I said to Mama one day, 'Mama, I'm gonna tell you something.' I was getting up big enough to really see needs. I said, 'If you don't start taking a little something and putting it somewhere so that if something happens to you, . . . there ain't gonna be nothing to bury us with.' I said, 'If you don't, I'm gonna start. I ain't never took nothing from you.' 'Cause when we were selling stuff, it was all turned over to her. I said, 'I'm going to start taking something and I'm gonna put a little bit somewhere.' So, Mama started [setting aside some money]." In the years that followed, Julia Benton Stokes provided her own "nest egg" and realized a degree of economic autonomy.[57]

Women's trade might even provide a way for them to achieve control over their reproduction. In the 1940s, Watauga County, North Carolina, residents participated in a pilot project that tested the efficacy of various birth control techniques. When rumors circulated that the public-health nurse planned to stop distributing free condoms, one woman "told her husband that she would simply catch a hen and take it to the store and sell it in order to get money to buy some Trojans." While many of the Watauga mothers did not have the fifty cents needed to purchase a package of three prophylactics, the nurse noted, she was heartened that "some will be willing to sacrifice their chickens and eggs in order to stop babies from coming."[58] Swapping a setting of eggs for reproductive freedom was enterprising, indeed.

In addition to the cash and credit that women earned, the relationships that they developed with customers accrued value for them and their families. By the 1940s, Julia Stokes and her daughter Lurline generated about fifty dollars a month from the route of long-term customers that they established and maintained. Although Lurline described her sales as a sideline to the farm operation, her earnings surpassed the profits that she remembered from the 1941 tobacco crop, when dry weather diminished yields and their golden leaf brought only $417 at season's end. Sales of produce, she observed, "helped pay for this place" and sometimes financed the living and farming expenses advanced to the tenant who helped her. While Lurline and her mother made money, they also made "an avalanche of friends" and accumulated a fund of friendship. Town women to whom they sold gave Lurline booties and baby clothes for her boys, items that she had no time to make because she was busy working on the farm. Contacts made among professional people in Florence, South Carolina, paid off in services rendered for reduced prices or other forms of help. When the family needed an electric motor but found it difficult to obtain, Lurline Murray consulted a furniture store owner who bought her buttermilk and eggs, and he located the equipment they desired. When the Murrays wanted to refinance their farm mortgage, they consulted a banker to whom she sold produce. Her mother's doctor, another longtime customer, refused to accept payments for Julia Stokes's care. "So often when I'd take Mama in there, Dr. Meade would say, 'No charge, Lurline. No charge.'" When she protested his generosity, he retorted, "'Now you just let me run my office.'" Lurline Murray concluded, "There ain't many people in the town of Florence that I didn't know. As I've said, I've sold to the richest and the poorest in the city of Florence."[59]

Marketing women such as Lurline Murray created a web of social relationships that included patrons of all classes. In fact, week in and week out, rural women sellers may have interacted with well-to-do people on more intimate terms than their men folks did. Men approached bankers, lawyers, and businessmen as borrowers and clients; market women dealt with them or their wives as entrepreneurs and providers of goods and services. Certainly, Lurline Murray's relationships with a variety of townsfolk expanded the family's network of contacts outside of the immediate neighborhood and gave her access to information about available services that could be enlisted in times of need. Perhaps because women's dependency on others seemed "natural," women could ask for help more easily than men, who feared losing face. Rather than imitating how town women spoke and dressed, rural women who sold at home demonstration curb markets were getting to know and establishing reputations among a town's most influential citizens and making friends among fellow marketers from all over the county. As Tom Cunningham noted, when his family moved it was his mother who helped the family step right into "the swing of things" and gained access to members of a new community by continuing familiar patterns at a curb market in a new place.[60]

Southern farm women, then, proved adept at producing and selling a variety of goods and pursuing a range of marketing strategies. Delivering eggs or butter to a local merchant, trading with a huckster, selling at the curb market, or supplying individual customers was but the first step in a complicated trade network. Farm women were crucial players in their household and local economies; perhaps unbeknownst to them, their goods entered regional and national economic channels. When women took what they had and turned it into money, their earnings kept families afloat during hard times and helped them enjoy some of the stock on store shelves during better times. Earnings might also shift the balance of power within families, giving women more influence over farm decisions and enhancing their sense of dignity and personal worth. In hindsight, it is clear that commodities like tobacco and cotton were the bricks of the southern farm economy, but the products of women's labor were often the mortar that held it together.

CHAPTER THREE

The Chicken Business

IN 1928 VANONA PATTERSON began searching for ways to increase the income of her family's farm in the North Carolina foothills. She wanted to send twin sons to college, and she worried that profits from tobacco would not pay tuition bills. Inspired by an article in a farm magazine, she tried selling pecans. Although she did "make some money" off the nuts, the earnings still would not underwrite her aspirations for her boys. "They just would never get an education if it had to come from what we was making," Patterson concluded. "So I went into the chicken business." Already a veteran grower of chickens that ran "in the yard," she turned to her mother-in-law to borrow the capital she needed to expand and improve her flock. "I went over to Grandma's to see if she would loan me some [money]," Patterson recalled. "She said she would. She gave me fifty dollars and that's what I bought my chickens with. Dad [my husband] didn't buy my chickens. I could get started with that much." To repay the debt, she later sold some roosters.[1]

When Patterson needed to learn about caring for a larger flock, she sought advice from the Alexander County extension agent and the vocational agriculture teacher at the high school. The agent and the teacher "had to learn, too," she noted. "I wasn't a-learning by myself. There wasn't no poultry business. People just kept a few chickens way back yonder, if you want to go way back." The chickens she raised produced high-quality eggs sold at the premium price of forty cents a dozen to a commercial hatchery about thirty miles away in Statesville and cull eggs that brought a penny apiece "on the open market." She always had a few live chickens for sale as well. Eventually, Patterson built four poultry houses that held some two hundred chickens each. "We'd build one house," she explained, "and the next year another one." Patterson managed her poultry business, deciding how to reinvest profits; she provided the labor, hauling feed from the granary and filling water troughs. "There was work to chickens then," she observed. "Work, plenty of it."[2]

Vanona Patterson's work paid off. Both of her boys graduated from North Carolina State College and landed jobs with federal agriculture agencies. One of those sons was visiting his mother during an interview on a spring afternoon in 1987. Rather than acting as memory prompter for his ninety-five-year-old mother as children so often do, he acted as a memory monitor who contested his mother's version of the role she played in running the farm and initiating changes in the crop and livestock mix. He and his siblings, the son reminded his mother, helped feed and water the chickens and shoveled manure from the poultry houses. And besides, his father owned the land where the houses stood and feed grain was planted. Although one hesitates to conjecture about the complicated tensions within the family, the son appeared to begrudge the credit that his mother claimed for the farm's success—and for his own.

Vanona Patterson and southern farm women like her were part of a quiet revolution in the region's poultry industry. As they experimented with new ways to raise chickens and demonstrated that their flocks made money, women helped build the foundation for the agribusiness poultry industry that emerged after World War II. During the first four decades of the twentieth century, poultry production increased in the South because women took advantage of a growing demand for eggs and fowls, new ways to produce them, and new ways to get those products to market. Farm educators found many women to be eager students of poultry-raising innovations. In the 1920s, state departments of agriculture and commission merchants sponsored railroad car-lot shipments of live chickens and eggs in an effort to centralize a complicated marketing chain. Women formed a vanguard as a fledgling poultry industry evolved, but their role has been overlooked.

Considering that poultry products added billions of dollars to the farm economy of the South by the late twentieth century, historians have given the industry's development surprisingly short shrift. Accounts of poultry raising jump from descriptions of the modest backyard flocks that women commonly kept to the full-blown vertical integration of the industry after World War II, with growers who, like latter-day sharecroppers, raised large flocks of broilers on contract for corporations that controlled the operation from hatching to processing. By telescoping its chronology, the story of the poultry industry turns into a predictable, linear narrative that pivots on a moment in which women and a part of the farm operation that they controlled lost out to science, capital, agricultural experts, and the men in their own families.[3] This truncated plot summary elides a crucial interim stage when women ruled the roost in the henhouse even as men jostled them for control.

Historians who have examined the commercialization of dairying and other "women's products" in other times and places suggest a more nuanced line of argument. In her study of the Philadelphia hinterland during the early nineteenth century, Joan Jensen found farm women increasing butter production for market when the price of grain collapsed, thereby providing their families with a steady source of income. Household dairying, Jensen concluded, "had in fact changed dramatically long before men entered into the occupation. Women developed butter making for the market, adopted the necessary skills and technology to increase production," and managed their own labor and that of children and hired workers. "In so doing women made possible the profitable commercialization of butter making." In a similar fashion, Sally McMurry focused on nineteenth-century cheese making families in upstate New York and how gender relations shaped the transition from domestic to factory production. While a male-initiated critique of women's cheese-making skills provided part of the ideological justification for the shift, women cheese makers themselves also had their own reasons for supporting the new arrangement. Tired of shouldering a workload that grew more burdensome as the scale of production increased and that seemed unfair when compared to men's labors, women abandoned home cheese production with little protest or regret when the opportunity arose.[4]

In poultry, as in the dairy industry, gender relations shaped the transition from domestic to commercial production. During the first half of the twentieth century, new research, technological changes, and growing markets paved the way for poultry expansion in the region. Women made the most of new possibilities, and as the story of innovators like Vanona Patterson suggests, allocating farm resources to poultry enterprises could foster power struggles that men and women had to negotiate.

By the time Vanona Patterson went into the chicken business, farm educators had been promoting increased poultry production as a new source of income for southern farmers for at least three decades. Poultry advocates recognized that women oversaw the care of flocks, worried that men ignored the value of fowls, and acknowledged that poultry inspired gender politics. In 1900, for example, the editors of the North Carolina State Board of Agriculture's *Bulletin* devoted the February issue to poultry. They intended their discussion "to be purely practical and helpful to the average housewife" in the state. After noting that "the good wife is at the head and front of the poultry department on most of our farms," the writers reprimanded husbands who failed to provide housing and feed for fowls even as they enjoyed eating fresh eggs and "juicy young chicken" and "pocket[ed] the money the good wife earns by patient labor and care of the home flock." "Well, well, what shall be said of" husbands who appropriated the profits of their wives' labor? the editors chided. But the writers skirted an answer in favor of helping "the housewife, and the husband, too, to a better understanding of the subject, and how to get the best results from the flock." Meanwhile, the editors noted that the most recent census had estimated that the chickens, turkeys, geese, ducks, and eggs raised and produced on Tar Heel farms were worth more than $3 million. To underscore the value of poultry, the editors observed that the figure exceeded by $400,000 "the total taxes on property collected in North Carolina for State, county and school purposes" in 1897. Moreover, the national earnings from poultry ranked third behind those from corn and cattle, yet "many men still laugh at the 'women folks' for their interest in the subject." If "farmers will give more consideration to poultry themselves," *Bulletin* writers concluded, "they will find larger profits in that direction than from any other source in proportion to capital invested."[5]

The *Bulletin* editors suggested that poultry remained underdeveloped because it carried the stigma of being "women's work." This assumption remained prevalent for years to come. In 1917 the extension farm agent in Union County, North Carolina, noted that he ran "the risk of losing prestige with the farmers" when be began to advocate better care of poultry. "The idea of there being any profit in poultry was ridiculed by almost every farmer," he declared. "The old hen was looked upon as a necessary evil that had to be borne to appease the housewife." Another Tar Heel farm agent claimed in 1933 that "poultry is looked upon on most farms as being an unnecessary nuisance." Some women suggested that their husbands hindered their quest for poultry success. In 1925 a North Carolina home demonstration club member described how she had started raising Plymouth Rocks "with no equipment, a damp uncomfortable poultry house and trees for their sleeping quarters." But she "wouldn't let herself get disappointed or discouraged." After she had convinced her husband that purebred poultry paid, he built a new henhouse and shared her enthusiasm. The husband had been "converted," and the wife's flock was now worth over $700.[6]

Compared to that in other regions of the country, poultry production in the South trailed considerably. At the turn of the twentieth century, for example, nearly 88 percent of North Carolina farms had poultry flocks, but they averaged just twenty-two birds each. Farm women in the upper Midwest, on the other hand, kept flocks that averaged sixty-one chickens, earning the region the moniker of the nation's "egg basket." Poultry and grain production went hand in hand in the heartland; railroads linked producers to buyers in Chicago and eastern markets. In a few pockets of the country, primarily near northeastern cities, some growers were beginning to specialize in the commercial production of broilers for meat. But in general, the availability of chicken was seasonal, highest in the late spring and early summer when poultry growers sold roosters hatched earlier in the year and kept pullets for laying eggs.[7]

By the early twentieth century, several technological advances set the stage for expanded poultry production. Two devices offered poultry growers more control over when eggs hatched and the care of young chicks. Farmers who used artificially heated incubators—sometimes called "wooden hens"—could hatch eggs at any time of the year, whether or not a hen was inclined to set on eggs and warm them with her plumage for the three weeks that it took an embryo to mature. Mechanical brooders kept newly hatched biddies warm and improved their chances of survival during the precarious first days of life when they were most vulnerable to cold. For growers who wanted to forgo incubating eggs and brooding biddies altogether, commercial hatcheries produced thousands of chicks that could be delivered to customers within hours by mail.[8]

Other developments at the turn of the twentieth century supported poultry expansion. Changes in transportation technology put scattered growers into closer contact with buyers in large urban markets. Railroad freight cars designed to hold hundreds of chickens and ventilated to keep the feathered birds cool during transit made hauling live poultry from farms to distant cities feasible and profitable.[9] Mechanical refrigeration lengthened egg storage time. Cases that held thirty dozen eggs replaced barrels, making it easier to handle large quantities of the fragile cargo.[10]

At the same time, professors at land-grant colleges and researchers at agricultural experiment stations devoted more attention to poultry husbandry and added poultry courses to their curricula. Researchers investigated ways to prevent and cure diseases that preyed on chickens and might destroy a flock overnight, and they advised growers to adopt one breed of pure-blooded birds. They designed houses that protected chickens from cold and damp weather, and they tested feeding formulas and regimens that enhanced egg production and broiler growth.[11]

Farm periodicals popularized poultry research and promoted poultry production. During the first two decades of the twentieth century, an editor with the folksy pen name of "Uncle Jo" presided over the *Progressive Farmer*'s "Poultry Yard" column. Mrs. J. C. Deaton of Rowan County, North Carolina, was a featured contributor, and women readers added to the store of knowledge by sharing lessons learned through trial and error. Besides a weekly column, in 1906 the *Progressive Farmer* introduced annual poultry editions that encouraged better care of fowls and smarter marketing of their products. "Diversify," the periodical urged readers in 1909, "make it cotton, corn and chickens and you will be starting on the way to 'easy street.'"[12]

Poultry topics usually fell into one of five categories: breeding, housing, feeding, preventing diseases, or marketing. "Uncle Jo" encouraged farmers to raise pure-blooded birds rather than falling "victim to the Southern mania for crossing breeds." For higher egg production during winter when prices peaked, the columnist advised growers to provide draft-free houses with a southern exposure rather than allowing chickens to roost in trees or build nests and lay eggs wherever they liked. Fowls also grew better when fed small grain and "green feed," such as young oats and rye, rather than the usual cracked corn and whatever they could find while ranging around the barnyard. Growers had to watch vigilantly for signs of mites, lice, and internal parasites that could cripple and kill chickens; cleaning nests and dusting hens with a caustic insect powder deterred the pests. Finally, eggs that were fresh, clean, and the same size brought the highest prices.[13]

Subscribers traded advice and testified to the profitability of their poultry flocks. In 1909 Mollie Tugman, a teenager from Watauga County, North Carolina, described the rigorous poultry care routine she had adopted despite her family's initial skepticism. When family members ridiculed her belief "that the poultry on our farm could be made to pay bigger profits," she solicited "the help of an eleven-year-old brother and a yoke of yearling oxen" and "sawed, dragged and set posts for enclosing a yard." Tugman then bought "palings and as the work proceeded, one after another of the family fell in line. Father and the boys dragged logs and had lumber sawed for a house and helped to build it." Next, the young woman "turned farmer and raised small grain" for feed instead of corn, which she considered "poor in egg-producing elements." As a result of better care, Mollie Tugman's hens laid so many eggs that when prices dropped she preserved them in a chemical solution known as water glass until early winter when they commanded "the highest market price."[14]

Women kept editors and fellow readers informed about the success of their flocks and revealed how they used their earnings. Mrs. H. P. McPherson of Moore County, North Carolina, emphasized that poultry income bestowed an independence that she valued highly. "Farmers' wives and daughters," she observed in 1907, "generally have a busy life, and no matter how generous and considerate the husband and father may be, there is always a feeling of dependence unless we have some way to make a little money of our own to spend as we please." Although she sold milk, butter, and vegetables, she found "more profit in poultry raising than in anything else."[15] A year later McPherson wondered, "How many of our good women do without many little things they need, because they are not always sure their husbands have the money to spare to buy them? Now there is little excuse for a woman to be without money if she has room to raise poultry." After recording the expenses and proceeds from poultry sales, she netted from $3.50 to $6 a hen during March and April.[16] In 1915 McPherson again reported that her hens paid "much better" than the cows she kept. "I do not keep many fowls," she wrote, "as I am not physically able to care for a large number, but I usually sell about $6 worth of eggs and chicks a year from each hen, and sometimes more. One season I sold $550 worth of chicks and eggs from sixty hens, and made most of my sales by advertising in the *Progressive Farmer*."[17] McPherson's experience seemed to fulfill the promise of a caption for a photograph of chickens in the magazine's 1911 poultry edition: "A woman needn't give up and think there is no way for her to make money because she is on a farm."[18]

The sale of fowls allowed some women to meet basic farm expenses. In 1918 a Virginia correspondent told "Poultry Yard" readers that her "aim with poultry has always been to make enough to pay the taxes." Because "poultry was usually selling high at tax time, from 1909 to 1916," wrote Mrs. H. T. Meriam, "I always had enough and a little over. Two years I sold a little over $164 worth, and those years they helped out on interest, too."[19]

Women used poultry profits to benefit their communities as well as their families. In 1914 a Tennessee woman described how her Cooperative Ladies' Aid Society helped pay off the church debt when the offering began to wane. Members started the Sunday egg project, agreeing to donate all proceeds from the sale of eggs that their hens laid on the Sabbath. "Then in order to get as much as possible for these Sunday eggs," Elizabeth D. Abernathy explained, "we thought of shipping them by the case to city markets." The church women asked friends who lived in Nashville, Montgomery, and St. Louis to serve as local retailers. Before long, members of the Ladies' Aid Society were sending five to eight cases—or 150 to 240 dozen eggs—each week, and they were selling "every-day eggs" as well as those laid on Sunday. With the women's donations, church members retired the debt and had enough money left over to repair and paint the building and furnish the sanctuary with a new organ. Not only did the "Sunday eggs" prove profitable, but also the project was one that women of all classes could join. Members "owning neither houses nor lands" contributed amounts from $2 to $12 from their Sunday eggs, Abernathy noted. "One of these women said to me: 'This has been the happiest year of my life. Working together is so nice. It has been such a help to me.'"[20]

Women who shared their poultry-raising methods in the *Progressive Farmer* maintained familiar practices even as they incorporated new ones. Neighboring women, for example, improved their stock by swapping settings of eggs from breeds or particular birds they considered superior instead of buying them.[21] They exchanged information on poultry care and ways to discourage predators. "Suck-egg" dogs that raided nests proved particularly vexing; forcing hot eggs into their mouths negatively reinforced the bad habit and foiled the canine menace.[22] Women who could not afford brooders to keep young chicks warm might solve the problem by bringing the biddies right into their houses and letting them hover in a cardboard box by the fireplace or stove.[23] *Progressive Farmer* readers offered cheaper alternatives to methods that the periodical endorsed. Although poultry column regular Mrs. Deaton considered a little beef scrap an "indispensable" feed supplement for good layers, a Virginia woman who had no beef to spare discovered that trimmings from fried pork worked just as well.[24] A Mississippi woman described constructing coops from dry-goods boxes and leftover lumber.[25] When a reader asked "Uncle Jo" how to prevent hawks from catching baby chicks, he first suggested deterrents that required store-bought ingredients. Then he shared a solution from a Florida "lady" who had heard it from "an old colored woman." Readers interested in following the technique should take "the shells which the chickens came out of, put them in a sack and hang the sack near where the chicks ranged, [and] the hawks would not bother the biddies." The Florida woman was "trying it this season, and up to now has not lost a chick. She said in her letter that I could laugh all I pleased, but it sure was a winner. It's cheap at any rate."[26]

Even without formal agriculture training, women experimented with methods and supplies. Mrs. H. T. Meriam, for example, devised her own hybrid techniques to raise geese, turkeys, and chickens. In 1916, she had improved her operation, buying an incubator that held 150 eggs, feeding new hatches rolled oats, and building a henhouse with a gravel floor. Even with the new equipment, however, Meriam used older methods for hatching eggs: she kept eggs warm in the incubator until "geese, turkeys or hens wanted to sit" and then let nature take its course.[27]

Farm women, then, could pick and choose among innovations in poultry raising as money and inclination allowed; they could adopt methods of improved housing, breeding, and disease control selectively, and without spending a lot. Regardless of the size or sophistication of their enterprises, women poultry growers could take advantage of expanding markets in the region's growing towns and cities. In sum, women were chief beneficiaries of developments in the poultry industry. By the 1920s, according to one chronicler of the industry, "farmers embraced poultry production with open arms. Or perhaps it would be more accurate to say the farmer's wife had found a new source of money—egg money."[28] An even more accurate assessment may be that farm women had discovered ways to reap higher rewards from a familiar source of cash and credit.

Women's poultry enterprises proved profitable. In North Carolina, for example, in 1909 some 4.6 million fowls sold for about $1.4 million, and over 10 million dozen eggs sold for $1.9 million. A decade later, North Carolina farms sold 3.1 million chickens and grossed over $2.1 million; some 11 million dozen eggs brought in $4.5 million. In 1929, the 5 million chickens sold off North Carolina farms brought in nearly $4.4 million, and the 20 million dozen eggs sold reaped some $6.3 million. All the while, one-half or more of the total amount of poultry raised and eggs produced was reserved for use at home, providing a valuable source of protein when farm families sat down at dinner tables.[29] In fact, a 1929 survey of Tar Heel farms revealed that between 1900 and 1925 the "poultry industry has developed more rap-idly . . . than has any other livestock enterprise." Women were in charge of a commodity whose income growth rate was a remarkable 221 percent.[30]

Enthusiasm for poultry quickened in the South in the 1920s, when boll weevils devoured the region's cotton and the price of all crops began to fall. While eggs and live chickens also responded to the market's ebb and flow, they remained more buoyant than most commodities. "Poultry products head the prosperity procession," the *Progressive Farmer* would crow in 1933. "They are the only farm commodities selling above pre-war prices."[31] Mean-while, the magazine championed poultry even more aggressively, and erudite college professors and researchers replaced folksy "Uncle Jo" as poultry col-umn editors. At the same time, specialized poultry periodicals joined the re-gion's general farm journals, and home and farm extension agents took the latest information about poultry directly to growers. Farm men began to join women in the poultry yards, but they had not yet displaced them.

In fact, poultry advocates displayed some confusion about the sex of their audiences. Some writers seemed to assume that larger flocks and improved care rested upon decisions about breeds and housing that only farm men would make. In the *Progressive Farmer*'s 1923 poultry issue, for example, F. J. Rothpletz addressed men when he recommended that they keep the poultry yard "reasonably convenient to the home and accessible in stormy weather" out of consideration for the women and children who did most of the work. At the same time, Rothpletz conceded "many women beat the men in poul-try management." And captioning the special edition's cover photograph of a flock of robust chickens was the statement attributed to "a good Southern farm woman": "The best way of making money I have tried is by raising poultry."[32]

Publishers of the *Dixie Poultry Journal* recognized women as linchpins in poultry production. Begun in 1921, the Nashville-based magazine spot-lighted women's enterprises, used their images to grace monthly covers, and featured an advice column written by Mary Fanning, a former teacher who had quit the classroom in 1915 to start a poultry operation that boasted prize-winning layers.[33] Advertisers of poultry feeds and supplies directed their appeals to female readers. Women described how they bred, fed, and

housed their flocks, marketed their products, spent their profits, and overcame adversity when predators or storms ravaged their broods.

"I think I must have been born with a mania for chickens," confessed Mrs.
Elizabeth Vogle, who credited her poultry earnings in 1926 to "just love and
care."[34] Other women outlined more meticulous poultry regimens that combined expert advice with makeshift solutions. Although Sallie P. West lobbied *Dixie Poultry Journal* readers to adopt purebred chickens, she acknowledged that her "own experience [was] proof that 'the best is the cheapest' in
poultry, as well as in other things." She purchased chicks and spent about a
dollar each fall to seed a winter pasture in rye for her flock of some thirty
Brown Leghorn hens, but jerry-rigged housing suited her standards. "My
coops or pens," West noted, "are such as a half-invalid woman can 'trigger
up' from odds and ends of poultry netting, tar-paper roofing, and scraps of
lumber." She ruthlessly culled layers who "loaf[ed] on the job," and followed
an "egg a day keeps the hatchet away" philosophy.[35] In a similar fashion, Mrs.
John Hughey's eye to good breeds and feeds and aggressive winnowing of
poor layers had produced what a *Dixie Poultry Journal* writer judged "[o]ne of
the most outstanding poultry successes . . . attained with an ordinary family
flock." The Tennessee woman noted that keeping equipment costs low was
central to accumulating profits of $600 in 1926. "Hens may lay as many eggs
in a palace as they would in a cheap, convenient hen house," the writer concluded after inspecting Hughey's operation, "but they don't lay as many
profits."[36] Expert advice to the contrary, Mrs. D. O. Ross admitted that she
did not keep her turkey hens penned up during laying season. "Half the joy,
thrills, and recreation would be taken away," she rhapsodized in 1927, "if I
didn't climb to the cliff among the red bud, dogwood, and sweet-scented lo-
cust and wild grape blossoms to find a cunning nest of large, speckled eggs,
well covered with leaves, to hide them from the crows." Ross assessed the
value of turkeys and eggs sold and the remaining breeding flock of 107 birds
at $700.[37]

Women revealed how they kept costs low and profits high. A *Dixie Poultry
Journal* reporter described the six-hundred-bird flock that a College Grove,
Tennessee, woman and her niece managed. "Their equipment is sufficient
but not expensive," Elizabeth L. Fowler wrote. The women had converted

an old barn "into as comfortable a laying house as the most particular of hens could desire." Discarded orange crates doubled as nests, and the duo fashioned their own feeding hoppers rather than buying them. The women estimated that sales of their eggs and baby chicks would top $2,000 in 1927.[38] When a busy Georgia farm wife and mother wrote a prize-winning essay for an International Baby Chick Association contest in 1929, the *Dixie Poultry Journal* solicited the secrets of her success. Although Mrs. G. G. Adair purchased hatchery chicks, she minimized expenses by converting a dilapidated house on the farm she and her husband had bought two years earlier into a brooder house and by using lumber from another rundown building for a new laying house for her flock of five hundred birds. When she added the contest prize of $1,000 to the profit of $2.90 she realized from each of her hens in 1928, Adair had earned enough to pay for their farm.[39]

In addition to raising poultry, women claimed a place in the auxiliary businesses that supported the emerging industry. The Statesville, North Carolina, hatchery to which Vanona Patterson sold her best eggs was owned and operated by Mrs. F. B. Bunch. The wife of a textile manufacturer, Bunch began incubating eggs in 1922 when she discovered that the local post office handled more than 80,000 mail-order chicks each year. Eager to tap the local market, she invested in incubators whose capacity reached 47,000 eggs in 1927, 80,000 eggs three years later, and 100,000 eggs by 1936. Besides relying on the help of her eight children, the busy mother valued the labor of an orphaned girl whom she had adopted. Teenager Annie Fink proved a quick study. She had built a laying house "with her own hands," and a *Dixie Poultry Journal* reporter thought that her "knowledge of poultry, balanced mashes, proper housing, [and] disease control" would rival "that of some of the best poultry authorities in the state."[40]

Still, when North Carolina women like Annie Fink and Vanona Patterson wanted poultry advice, they could count on farm extension agents and male agriculture teachers as well as home demonstration agents. In 1926, 360 Iredell County men and women attended A. R. Morrow's lessons on culling flocks of poor layers, proper feeding, and housing.[41] In his 1926 annual report, agent J. T. Monroe of Jones County included a photograph of "interested poultrymen & poultrywomen" who had attended a demonstration at

"Poultrywoman and poultry specialist going over records. Mrs. Bunch & Mr. Parrish, North Carolina, May 1930." (S-13723-C, Record Group 16-G, U.S. Department of Agriculture, Box 59, Animals-Chickens-Marketing folder, National Archives and Records Administration, College Park, Md.)

the home of Mrs. J. R. Westbrook. Westbrook had increased her flock from five hundred to eight hundred White Leghorns, and had realized a profit of $1.50 per hen. In addition, after installing an incubator with a 4,600-egg capacity, she hatched eggs for neighbors and "was rushed the entire hatching season."[42] That same year the agent in Madison County described the operation of a premier poultry grower, Mrs. J. E. Bryan, who kept a flock of Barred Plymouth Rocks and had invested in an incubator, brooder, and "an up-to-date hen house," and praised the work of Mrs. Otis Chandley, "a real poultry lady."[43] The home demonstration meeting that drew the largest crowd in Guilford County in 1930 featured a vocational agriculture teacher discussing poultry diseases. The fifty-four women who attended "kept [him] quite a while answering questions."[44]

Besides demonstrating poultry care, during the 1920s agents of state governments turned increased attention to marketing fowls and eggs and helping growers sell their products beyond their communities. Poultry promoters worried that farmers failed to understand the demand for chickens and eggs and their value if they had sold only in local markets that were flooded during seasons when production peaked. "Markets are not wanting for any poultry produce," the *Dixie Poultry Journal* assured readers. "The one thing needful is a way to get the products to market."[45] Yet when one industry observer contemplated the "multitude of channels" that eggs might follow from producer to consumer, he concluded that there "are so many combinations of circumstances involved in marketing eggs that no attempt can be made to describe all of them in detail."[46] While some producers established egg routes or used parcel post to sell directly to customers, others cultivated institutional buyers, such as hotels and restaurants. Most traded or sold eggs and live chickens to country storekeepers or traveling hucksters, who in turn wholesaled poultry products to shippers and jobbers.[47]

Selling cooperatively and tapping markets beyond local communities, agriculture officials advised, would help producers reap greater profits and encourage higher production. In 1921 a home demonstration agent in Murfreesboro, Tennessee, helped "a handful of women, interested in making poultry pay," form an egg circle. The agent taught the charter members of the Rutherford County Poultry Association how to grade, clean, and case their eggs. During the first six months, the group appointed one member each week to receive their eggs and sell them to local merchants. The women then branched out, shipping their first three cases of eggs to Philadelphia, where merchants paid twenty cents a dozen more than hometown buyers. When word of the higher prices spread, the egg circle grew, and its members soon shipped fifty cases of eggs a week. Their appetite for profits whetted, the women decided to "establish their business more firmly." They raised stock valued at $1,400, secured a building near the cooperative creamery, solicited more members, and hired a man to manage their business. The manager issued checks bimonthly, and by 1926 the egg circle sold $40,000 worth of chickens and eggs. Many members were "astonished at the money this phase of farming is bringing in."[48]

In the 1920s the home agent in North Carolina's Anson County, Rosalind Redfearn, lined up buyers for fowls at colleges and engaged customers as far away as Washington, D.C. For several months in 1922, home demonstration club members shipped a barrel of dressed hens more than a hundred miles to State College every Friday, and in November they sent over a thousand pounds of turkeys to the Raleigh school. The price the college paid, Redfearn noted, was nearly double that of glutted local markets.[49] To prepare fowls for shipping, clubwomen brought their birds to a member's house, where they killed and plucked them and hung them to cool overnight. The next morning a member packed the birds in a barrel and took them to a railroad shipping point. "At these community 'hen parties,'" Redfearn observed, "the people had the best time."[50]

Departments of agriculture in southern states also organized poultry marketing cooperatives that shipped railroad car lots of eggs and chickens to buyers in New York and other distant markets. "The shipping of the first car of live poultry from a county was like a holiday for those participating," North Carolina agriculture officials reported in the 1920s, "and although there was much work back home to be done, small groups of farmers stood around all day and discussed the poultry car and what it might mean to poultry income. Farm women especially went back home fully determined that they would have more poultry for sale at the proper time and were thankful that they did not have to drive around all over town to sell a few old hens."[51]

In 1925 the home demonstration agent in Robeson County, North Carolina, helped inaugurate railroad shipments of poultry from local farms. The experiment proved a grand success; the sale of some 32,000 pounds of chickens grossed over $7,000, a sum that convinced farmers that poultry was one "cash crop" they could count on. On the day of the shipment, the county seat of Lumberton took on a festive spirit. The crowds grew so large that the agent "had to call on the police force to help handle traffic. Cars, wagons and buggies of every kind were parked along the street three blocks from the [railroad] car door, all waiting to be served." Out of curiosity, peo-

First cooperative poultry and egg shipments, North Carolina. (N53.16.1003, Archives and Records Section, N.C. Division of Archives and History, Raleigh)

ple "came from far and near to see the big show," and the "merchants opened their eyes and the stores were as busy as on tobacco sales days. Many said it could not be done, but all stood by and the car was packed beyond a question."[52]

When observers alluded to a holiday atmosphere to describe wholesale poultry marketing, they invoked a sense of carnival. In much the same way that carnivals historically invited gender inversions and temporarily transposed relations of power, now a commodity traditionally controlled by women was grossing impressive sums and giving other farm commodities a run for their money. Poultry shipments rivaled tobacco markets for generating cash and stimulating commerce. Farm women's financial contributions could hardly be ignored.

During the depression, North Carolina extension agents described a topsy-turvy economy in which poultry often maintained farms and families. In Beaufort County, farm agent E. P. Welch reported in 1931, poultry was "keeping the farm in operation in numbers of cases. The poultry rank along with the gardens in enabling the small farmer to have something to eat and some spending money."[53] Farm and home agents all over the state echoed these findings. In Forsyth County, poultry was "the one bright spot in the great Agricultural structure of our county."[54] In 1932 agent Lillie H. Hester observed that "poultry money has meant much to the farm homes of Bladen County" and described how one woman's flock supported a family of nine.[55] Two years later Hester reported that "the poultry has helped to pay the taxes, buy books and clothes to keep the children in school."[56] In 1931 the farm agent in Carteret County related how a tenant woman had raised and sold two hundred turkeys to good advantage. "She paid the family fertilizer bill and grocery bill," Hugh Overstreet commented wryly, "for her husband to produce five cent tobacco."[57] That same year, in Durham County, Mrs. E. A. Perry had earned $206 selling fryers and broilers, or "three times as much money from her chickens as Mr. Perry made from his tobacco. A new brood house is now under construction and both Mr. and Mrs. Perry will raise chickens next year."[58]

In 1933 the *Progressive Farmer* boasted that the southern poultry industry had "attained the magnitude of big business" and been transformed from

"A 4-H club member feeding her poultry flock, which has 65 laying hens." (s-6777, Madison County, Fla., 1941 [?], U.S. Department of Agriculture Extension Service, Library of Congress, Washington, D.C.)

"an unimportant farm chore—throwing out a little corn and collecting a few eggs—to a scientific business and a major source of farm income." Although there were fewer chickens on southern farms in 1930 than in 1920, the cash value of the eggs had nearly doubled—from some $108 million in 1924 to $199 million in 1929. And in 1931 more than 10,000 railroad cars of live poultry had rolled from the South to markets in New York, Boston, Philadelphia, and Chicago. "Who said Dixie's chickens aren't in the scientific big business class?" a writer asked.[59]

The *Progressive Farmer* continued to encourage women to act as agents of change and acknowledged that their poultry-raising talents often surpassed those of men. In March 1935, C. S. Eastridge wrote to the magazine charging that poultry generated no profits. In turn, the *Progressive Farmer* invited subscribers to respond, offering eight cash prizes valued at $2 to $15 for the best "experience letters." Of the 802 correspondents who replied, only one agreed with Eastridge. "The other 801 said he was wrong," an editor reported. "Their explanations of his failure ranged all the way from a suggestion that his 26 chickens were all roosters to one which thought he probably had an egg-sucking dog . . . and another who said his trouble must be that he left his chickens to the 'wimmen folks.' But the latter suggestion is ruled out because 'wimmen folks' appear to be more successful than men with poultry." The first-prize letter came from Opal McCain of Clay County, Alabama, who used "the best knowledge science can offer" to make chickens pay. "The year I finished high school," McCain wrote, "I selected poultry as a means of making a living." Between March of 1933 and January of 1935, she had earned $472 from her flock. McCain credited her success to "beginning in a scientific way" and following the advice of a variety of agriculture professionals.[60]

Poultry promoters reserved their most exuberant praise for growers who, like Opal McCain and home demonstration clubwomen, adopted the practices that the "experts" preached. They did not assume that women, science, and "big business" were incompatible; on the contrary, they admitted that women could adopt new methods and keep records of profit and loss as well as—or better than—men. When poultry advocates condemned traditional methods of raising poultry, their criticism was not necessarily targeted at women and appears to have been no harsher than rebukes reserved for farm men who failed to follow the lessons of scientific agriculture.[61]

While some farm women heeded the suggestions of county agents and poultry "experts," others sought their opinions only occasionally and continued to adopt new methods selectively. Lurline Stokes Murray of Florence, South Carolina, fell into this category. Rather than raising one breed, she preferred "a duke's mixture" that included several types of birds. Scrap lumber and sticks used to hang tobacco in curing barns provided building materials for the pens and houses that protected her chickens. But when she needed to make water troughs, she followed instructions provided by the extension service, and she called upon extension agents and hatchery workers when a poultry disease stumped her.[62] A *Progressive Farmer* subscriber identified as a

Lurline Stokes Murray gathering eggs, Florence, S.C., June 1987. (Photo by Eric Long, 87-16467-2, Office of Imaging, Printing, and Photographic Services, Smithsonian Institution, Washington, D.C.)

"Farm Woman" from Madison County, Georgia, told readers how she made money on broilers even though she could not afford to buy an incubator. She had fashioned coops from boxes draped with sacks; she fed the chickens cracked wheat and corn, table scraps, and buttermilk. The woman concluded, "It doesn't take fine houses or expensive equipment to raise chickens."[63]

By the 1920s farm men were beginning to enter the poultry business, especially in the upland South where cotton farming had long been marginal. Arthur Fleming was born in one of those hill country areas in 1917, near the north Georgia town of Gainesville—now the self-proclaimed poultry capital of the world. The poultry industry was in its infancy when Fleming was growing up with his grandparents after being orphaned as a young boy. His grandmother raised hens, swapping eggs for groceries at a nearby country store or selling to a local huckster who trucked country produce to Atlanta, about fifty miles away. "He got to where he'd come through and stop once a week," Fleming recalled. "'Mrs. Fleming, have you got any chickens? Have you got any eggs? Arthur, have you got any rabbits?'"[64] In 1923, Fleming's grandfather built a chicken house, raised his first batch of five hundred broilers, and hauled them in a two-horse wagon to Murphy Brothers. The commission merchants paid Fleming's grandfather on the spot and shipped the chickens by rail to Chicago and New York.[65]

When Fleming's grandfather started raising large flocks, his grandmother's yard flock had to go. Commercial growers worried that chickens that ranged freely would contaminate their birds and that diseases could spread rapidly among chickens raised in close quarters.[66] While individual women might lose control of a source of cash and credit, women as a whole were not yet being denied a piece of the poultry pie. And when men headed commercial poultry operations, they often depended upon the labor of wives and children.

Like farmers in north Georgia, poultry growers in the foothills of Wilkes County, North Carolina, began to experiment with larger flocks in the 1920s and 1930s. Gerti Roberts's mother expanded her operation after she was widowed in 1934. First, her daughter recalled, her mother raised layers that produced hatchery-quality eggs; later she switched to raising some two hundred broilers. The larger and improved flocks required better housing and feed than her yard flock had, so Roberts's mother built a wood-fired furnace out of rocks and cement to keep the chickens warm, and she bought feed from the hatchery in hundred-pound bags.[67]

Women like Roberts's mother were helping to create what North Carolina extension officials proclaimed as "one of the leading farm enterprises" by the late 1930s.[68] In 1938, the poultry extension specialist noted that the number of "semi-commercial" flocks of over fifty birds had grown by 23,000 that year and observed that Tar Heel farmers were "realizing that poultry is a regular revenue producer."[69] Women were responsible for much of the increased production. As late as 1940 the poultry specialist conceded that many men still deemed it "a disgrace to be caught in the chicken house" and therefore refused to invest in livestock whose care was assumed to be "a woman's job."[70]

Looking back on the development of the Wilkes County, North Carolina, poultry industry, pioneer entrepreneur Charles O. Lovette credited much of its early inspiration to farm women. According to Lovette, more than half of the first broiler houses "were built at the insistence of the farmer's wife," who in the 1930s already understood the economics and care of poultry. Once the chickens proved profitable, however, men might appropriate the operations.[71]

Charles Lovette himself had gotten his start in 1924 as a huckster who bought chickens, eggs, and vegetables from farm families and country stores in Wilkes County. Like other itinerant buyers, he covered specific routes on specific days, filling up the bed of a Model T Ford truck with country produce. As a youngster, T. H. Kilby accompanied Lovette and helped pack the eggs and catch the live chickens offered for sale by "the ladies . . . standing out on the side of the road." He "had a little pair of scales that he carried with him," Kilby recalled. "He would weigh the chickens and pay them for that." Lovette then peddled the products to stores, boarding houses, and restaurants in Charlotte and Winston-Salem, each less than a hundred miles away.[72]

By 1927 Lovette's business had grown enough to keep two trucks busy buying and selling eggs and live chickens, and his young wife became a quiet partner in the enterprise. Although Ruth Bumgarner Lovette had eight children between 1925 and 1946, she sometimes accompanied her husband on buying trips, often repacked produce in preparation for its resale in town, and regularly purchased goods that people brought to her back door. "I could never cook a meal or churn or do anything that I didn't have to quit and buy something," Ruth Lovette recalled years later. "Many people have asked me why I didn't learn to drive. I told 'em I was too busy having babies and buying chickens." While she and the older children sorted and packed produce for sale, Ruth Lovette discovered that an empty turkey coop provided the perfect playpen for the baby. At the end of the day, when Charlie Lovette returned from selling, the family "knocked off and went into the house to settle down to a good night's rest with the knowledge of a good day's work done. To me this was happiness, knowing that I was pulling my share of the load in making the living for our ever-growing family." But by the early 1940s, the business "was taking over our home life" and so its head-quarters shifted to the town of North Wilkesboro.[73]

In 1943, Fred Lovette joined his father's huckstering business. Two years later, with a loan from his father, he expanded the poultry buying and sell-ing business. In less than a decade, Fred Lovette had purchased a chicken processing plant and what would become North Carolina's premier poultry agribusiness, Holly Farms, was born.[74]

Fred Lovette entered the family business just as demand for eggs and chickens soared during World War II. In need of food for the troops, the federal government encouraged poultry production and bought large quan-tities to serve in military mess halls. The war provided farm men ideological justification and economic incentives for raising poultry. Farmers who fat-tened broilers and coddled good laying hens no longer suffered the ignominy of doing women's work. Instead, they now were "soldiers of the soil" who answered Uncle Sam's call to service. Patriotism paid off, and poultry grow-ers reaped good profits.[75] By 1943, North Carolina farmers had "shattered" all previous poultry records. Never before had Tar Heel farmers "raised so many chickens, produced so many broilers commercially, kept so many eggs" as that year. "Yes," the poultry extension specialist concluded breath-lessly, "it was truly a year in which all production records were smashed."[76] Wartime demand accelerated the evolution of the industry and fueled the rise of North Carolina and other southern states to the top of poultry pro-duction charts.

As poultry raising grew in scale and required more capital investment, farm men usually assumed the position of managers while women and chil-dren performed much of the actual work.[77] Women's loss of autonomy prefigured the erosion of independence that their men folks, in turn, would experience when they began growing broilers on contract with corporations like Holly Farms. Poultry agribusinesses closely supervised the operations of contractors, made key managerial decisions, and eventually reduced growers to little more than wageworkers.[78]

But in the first four decades of the twentieth century, how the poultry industry would mature into "poultry factories" in the South was still not determined. During those years, poultry raising passed through a stage in which farm women could expand their barnyard flocks, adopt new production methods, and take advantage of growing markets in towns and cities as money and inclination allowed. With a modest amount of capital, some women like Vanona Patterson "went into the chicken business" and proved that poultry could be profitable. In the process, women's operations attracted the attention of men who once might have greeted their poultry efforts with disdain.

CHAPTER FOUR

Professional Paradoxes

IN THE SPRING OF 1922, Charlotte Hilton Green, a magazine writer, attended a two-week training session for North Carolina's white home demonstration agents in Greensboro. She called the women she met "daughters of the New South." Green speculated that the agents had descended from the "powdered belle of long ago who trod a stately minuet." But "the days of southern belles with old black mammies to wait on them" were gone; in their place were "self-reliant young women" such as the home demonstration agents who accepted this "missionary work" in order "to develop their beloved Dixie." Green contrasted the dedicated, energetic young women touring along in their Fords with the country woman who "spends long, weary days in the field and kitchen, almost forgetting how to smile, until into her life comes the home demonstration agent with new suggestions for home helps, balanced meals, remodeling the wardrobe and refinishing old furniture." The farm woman, Green noted, "has been called 'the unpaid servant of the nation and the most indispensable member of Uncle Sam's big family,'" so it was only fair that the government should repay her labors by helping her market surplus produce, acquainting her with inexpensive household conveniences that would lighten work, and bringing "variety and recreation into the routine of her days."[1]

Among the "daughters of the New South" gathered in Greensboro was Anne Pauline Smith of Franklin County. A seasoned agent at the age of thirty-two, Smith had just won a promotion and was preparing to assume new duties as the district agent in charge of northeastern North Carolina. Smith's personal papers, comprised largely of letters exchanged with her longtime fiancé, offer insights into an agent's life. Rare in their candor, these letters trace the development of home demonstration from its beginnings as a reform movement to a profession with its own culture. They complicate agents' public image as selfless "missionaries" motivated primarily by service. The rhetoric of service could mask as much as it revealed; it muted the complex reasons that inspired women to pursue employment as a home agent and obscured the conditions of their work. For Pauline Smith, and no doubt for many of her peers, this occupation provided an arena in which she could influence public policy and win social recognition. It rewarded her with a salary that ensured economic independence, which in turn allowed her to consider marriage an option rather than a destiny.[2]

By helping us read between the lines of official documents and memoirs, Smith's letters reveal the pressured and politicized environment in which agents worked. Her correspondence underscores the personal and professional paradoxes that she and other agents had to negotiate. Ironically, despite her dislike of household chores, she made a career of teaching rural women how to be better housekeepers and community members, and her professional success hinged in large measure upon the approval of the very women whom she deemed in need of help. She promoted ideals of domestic life that the demands of her job made it impossible for her to achieve. As Smith became a new woman of the New South, she wrestled with old ideas about women's proper roles and envisioned a new man of the New South.[3]

FOR WHITE COLLEGE-EDUCATED women like Smith, government efforts to change the lives of farm women created job options and made them key players in a broad program of rural reform. A more prosperous countryside, agrarian progressives believed, depended upon rural women and girls' learning to apply scientific methods to familiar domestic chores and to market the products of their labor more aggressively. In 1914, Congress passed the Smith-Lever Act, establishing the extension service of the U.S. Department of Agriculture (USDA) and supporting the work of farm and home demonstration agents in cooperation with land-grant colleges.[4]

In North Carolina, the extension service built upon established efforts to reach rural women and girls. Since 1906, the state department of agriculture had sponsored Farmers' Institutes for Women. Then in 1911, contributions from the General Education Board (GEB), a Rockefeller philanthropy, allowed the state agriculture department to offer training in home economics for farm girls as it offered training in agriculture for men and boys. The first home demonstration agents organized Tomato Clubs where girls learned scientific methods for growing and canning tomatoes, marketed their harvests, and pocketed the proceeds. Tomato Clubs represented a modest start for what grew into 4-H Clubs for youth and home demonstration clubs for adult women. Home demonstration agents heralded the gospel of domestic reform and economic development in the rural New South.[5]

Because the job was a new one in the 1910s, qualifications for home demonstration agents were defined vaguely at first. When North Carolina's state leader, Jane Simpson McKimmon, sought Tomato Club organizers, for example, she looked for "women of good education, with a background of culture in the home," "a practical knowledge of homemaking," "good business ability," and a familiarity with rural life. Most early agents were single public school teachers; a few were married women like McKimmon who had distinguished themselves in their local town's women's clubs or as Farmers' Institute lecturers.[6]

As leader of North Carolina's home demonstration program, Jane McKimmon ranked among the most revered Tar Heel women during the first half of the twentieth century. Despite widespread public recognition, her private life remains sketchy. Her husband and children merited no more than a passing nod in her professional autobiography. It is as if McKimmon took pains to reveal only the aspects of her life that reinforced her image as a woman who simply answered a call to serve. This much we do know. Born in Raleigh in 1867, Jane was the oldest of Anna Cannon Shank and William Simpson's nine children. Her mother concentrated on raising the family, and her father, a pharmacist, ran his own drugstore. At the age of sixteen Jane graduated from Peace Institute, a private girl's junior college in her hometown. After her parents squelched her dreams of attending art school in New York, Jane became a wife and mother. In 1886, at the age of nineteen, she married a Raleigh merchant nearly twenty years her senior, Charles S. McKimmon, and they had a daughter and three sons. Like many middle-class women of her generation, she joined the Raleigh Woman's Club, where she combined self-development and civic improvement.[7]

McKimmon took her first paid job in 1909 when she joined the Farmers' Institutes for Women lecture circuit. Institute speakers traveled from county to county offering instruction on topics such as cooking, sanitation, and marketing. McKimmon's specialty was bread making. At the institutes farm women learned practical skills and, perhaps more important, found a new appreciation of their own domestic work and camaraderie in the company of other women. Years later McKimmon recalled one poor farm mother, in particular, who made a valiant effort to bring her daughter to a cooking class. The woman had disregarded her husband, who had mocked the yeast-roll demonstration as "all foolishness," and completed her fieldwork before day in order to arrive at the institute on time. "I told my old man I was a-coming," she explained, and such examples of heartfelt determination and domestic defiance convinced McKimmon of the value of her work. By 1911, McKimmon directed the women's institutes, earning a reputation as a good administrator. When the GEB donated money for rural girls' clubs that year, McKimmon became one of five pioneering home demonstration agents in the nation. Her first task was to recruit a summer staff.[8]

When McKimmon began hiring agents, she often turned to young women like Pauline Smith, who in the summer of 1913 was teaching the children of tobacco farmers in the Franklin County hamlet of Seven Paths. Smith, who had grown up in the county seat of Louisburg, belonged to a family where it was not unusual for a woman to work for wages. Her mother, who died in 1909, taught school before and after she married. Her father made his living by painting houses. As a teenager, Pauline worked as a substitute teacher at the Louisburg Graded School, on-the-job training that she supplemented with courses in pedagogy at the East Carolina Teachers Training School, which had just opened its doors in Greenville.[9]

Whether married or single, volunteers or wage earners, the pioneer generation of agents came of age during the heyday of reformist zeal in the South and the state. Members of women's voluntary associations and female teachers marched at the forefront of education enthusiasts who considered schools the foundation for a revitalized southern economy and culture. Fueled by a sense of mission, middle-class women expanded an ethos of service to sanction their entry into the public world of policy debates and paid employment as they encouraged taxpayers to support school construction and then staffed the thousands of new classrooms created during the first years of the twentieth century. Work as a home demonstration agent provided women a new venue in which to strive toward familiar goals.[10]

Years later, Ruth Evans Dozier reminisced about her experiences as a Tomato Club organizer. She was a single, twenty-eight-year-old schoolteacher in Pitt County when McKimmon invited her to supervise club girls during the summer of 1912. Dozier initially demurred, protesting that the job would preclude taking a vacation with her family. "But my dear," McKimmon replied, "think of what a wonderful service you could render to help others." Dozier succumbed to McKimmon's logic, and returned to the job for the next four summers.[11]

In the beginning McKimmon and her agents "learned by doing" as much as the Tomato Club girls did. Dozier reported they "made a big mistake," for example, when ten girls planted their one-tenth-acre plots all at once. "Can you imagine," Dozier mused, "how many tomatoes we had with the whole acre ripening at the same time?" In addition to learning by trial and error, agents studied modern canning methods at training schools held at women's colleges. Formal instruction boosted their confidence, but it could also put them at odds with rural women's customary practices. During a demonstration where she stressed "the need for sterilization," Dozier overheard a volunteer canning assistant stage whisper to a club girl, "'You don't have to do all that. Tomatoes will keep any way.'"[12]

During the summers of 1915 and 1916, Dozier acted as a roving agent, organizing girls' clubs across North Carolina from the mountains to the sea. An intrepid traveler, Dozier encountered new economic and cultural terrain as well as new places. Arriving by train, she might eat breakfast in one town, dinner in another, and supper in a third. "I wasn't afraid of anything then," Dozier recalled. "I'd go into a town, call a livery stable, ask that a horse and buggy be sent to my hotel that I want to drive out into the country." The conditions that she found in some rural homes startled her. While staying overnight with the family of a Mecklenburg County Tomato Club member, Dozier discovered that the bed and walls were crawling with bugs, so she spent a restless night nodding in a chair. "Oh, yes," Dozier lamented, "a spoiled, sheltered girl raised in a clean, comfortable home didn't know that such homes existed" in North Carolina.[13]

Common themes mark the reminiscences of pioneer agents like Dozier. They betray a voyeuristic fascination with people at once similar to and yet so different from themselves, and they wed service to a sense of daring. Travel escapades that combine extraordinary dedication to the job with an element of risk figure prominently in the chronicles of agents. They were mobile women who sometimes appeared to be out of place to men with whom they shared country roads. Before Mary Wigley Harper spent a year as an agent in North Carolina, she began her career in her native state of Alabama. The difficulties and dangers of traveling to club meetings lingered in her memory. On one occasion Harper was so determined to arrive on time at a meeting in a distant community that she began the journey by horse and buggy at two in the morning. When she encountered a car, Harper guided the horse to the side of the dark road for fear that bright headlights would spook the animal. While Harper waited for the car to pass, the vehicle slowed down and its male occupants yelled at her. "Nothing like it had happened in my travels before," she remarked. Shaken but unscathed, she forged ahead, arrived at the meeting on schedule, and completed her demonstration with her usual aplomb. On another occasion, however, Harper felt more threatened when she passed a car in which the male driver looked at her and leered, "Hello, sweetheart." When Harper reported the incident to the sheriff, "he said from now on I must be prepared to protect myself. He made me a deputy sheriff so I could legally carry a gun." Along with cooking utensils, she packed the weapon as a standard part of her equipment.[14]

Pioneer agents counted physical stamina, fearlessness, dedication to service, and an ingenious ability to solve problems as characteristics necessary for success. Yet those traits coalesced into stereotypes that obscured the reality and complexity of their professional lives. Jane McKimmon's cleverly titled professional memoir is a good case in point. *When We're Green We Grow* reads like an adventure story featuring agents as unflappable heroines who act out of a selfless devotion to duty. They fight muddy roads and brave broken automobile axles to arrive at meetings, where they conduct demonstrations to inspire skeptical farm women to perform routine tasks in new ways and use the gathering to brighten rural women's dreary lives. In her memoirs, McKimmon suppressed conflicts and reserved no place for agents who failed to fit the conventional mold. Her characters rarely develop or mature.[15]

Home agents had good reasons for creating a carefully constructed public persona. Most were young and single women who were newcomers to rural communities. Visible and mobile, agents had to be prepared for all aspects of their private lives—from how they styled their hair to the cut of their clothes and the company they kept—to be the subject of public scrutiny. Even the appearance of impropriety could sabotage their work. In 1921, for example, the Onslow County farm agent relayed word to his superiors that Margaret Martin, the home agent, had "gotten the ill will of most all the leading citizens"; they might even petition the commissioners to cut the appropriation for the position. Part of the opposition stemmed from Martin's "driving with a bunch of sports" in Jacksonville one Saturday afternoon when she was supposed to be demonstrating a fireless cooker. The farm agent believed that "the trouble [could] be prevented" and the position saved if another agent were appointed. Upon further investigation, however, Martin's supervisors pieced together a different story. While she *had* been involved in a relationship with a young man that raised some eyebrows the year before, the relationship had ended. Martin's failure to appear at the demonstration was simply a scheduling mix-up. It appeared that the farm agent was spreading rumors about her romantic liaisons. Hearsay or fact, Margaret Martin's credibility had been undermined, and the commissioners asked her to resign later that year.[16]

The smattering of correspondence that chronicles Margaret Martin's problems fractures home agents' public façade and moves beyond caricature, heroic or otherwise. In a similar fashion, the letters of Jane McKimmon and Pauline Smith suggest incongruities between the images they developed and the reality of their lives. Certainly, multiple motives guided their careers.

By all accounts, McKimmon was a beloved worker and an inspirational supervisor who drew agents under her wing. Admirers canonized her. When a writer for *Country Gentleman* magazine arrived at her office in 1918 to chronicle the success of the North Carolina program, she found that McKimmon's "face had something of the expression one sees in certain of the medieval saints of the executive type—spiritual, but outgoing and capable." This saint, the writer intuited, possessed "a keen sense for business opportunities and for a cultivation of the main chance—not for herself, but for those of whom she has charge." Pressed to explain her achievements against daunting odds, McKimmon quoted a favorite verse about a "little cork" that outmaneuvered an angry, tail-thrashing whale because its inherent buoyancy allowed it "to float instead of to drown."[17]

Behind the saintly demeanor, McKimmon faced private difficulties that required cork-like resilience. Decades later, Ruth Evans Dozier declared that McKimmon had gone to work because she "married a high society playboy who knew how to be a foolished [*sic*] gentleman . . . , but he didn't know how to make a living for his family." Dozier's assessment of Charles McKimmon's inadequacy as a provider is confirmed by a small collection of family letters written during World War I. Not long before the *Country Gentleman* writer profiled McKimmon, she and her husband had moved from their rambling Raleigh house into a small apartment rented in her name. Whether the couple chose this new living arrangement because all their children had left home or in response to financial difficulties is unclear. But Charles McKimmon's dry goods business had failed, and money matters definitely preoccupied the family. "Son," Jane McKimmon wrote in early 1918 to Billy, in army flight-training, "I think God is very good to make it possible for me

Jane Simpson McKimmon, North Carolina's state home demonstration agent, 1911–36. (Photo courtesy of North Carolina State University Archives, D. H. Hill Library, Raleigh)

to earn enough to keep father and me in such comfort." A few weeks later, Charles McKimmon needed "to talk a little business" with his son, and he encouraged Billy to repay a loan from his mother because "we are a little cramped." Billy complied and began sending a portion of his earnings to his parents each month.[18]

Her public pronouncements notwithstanding, Jane McKimmon acted upon more complex motivations than a humanitarian impulse. She needed money, and she drew deep personal satisfaction from her work. Although she demurred at the praise that federal and state agriculture officials bestowed on her and the program she directed, she nonetheless admitted to her children that she could hardly keep her head from swelling. "You see," she joked, "I am wearing a very large hat these days!"[19] McKimmon used her talents, enjoyed exercising authority, and emerged as a mover and a shaker.

In McKimmon, Pauline Smith found a mentor with whom she shared a desire for service and accomplishment and a history of men who fell short as providers. Her father, Smith recalled, "did well to be a drinking man," and the financial insecurity that his intemperance created inspired a deep desire for economic autonomy in his daughter. Although official accounts usually soft-pedaled economic incentives as a reason for women to pursue extension work, they played a central role in Smith's decision to abandon school teaching. During her first summer as a Tomato Club organizer, Smith learned a lesson as valuable as those that her charges absorbed by cultivating, canning, and selling tomatoes: work as a home demonstration agent proved more lucrative than the $55 a month that she earned teaching school. When McKimmon offered Smith a salary of $300 for the summer, she must have been as ripe for the picking as the fruit the girls were to grow. In 1918, McKimmon appointed Smith the permanent home agent in her native county.[20]

Like Smith, many early home demonstration agents began their working lives as teachers, but once they joined the extension service they formed a branch of the new female profession of home economics. Rooted in nineteenth-century domestic science and the more recent interest of women in chemistry and nutrition, home economists formed their own professional society, the American Home Economics Association, in 1909, and instituted formal curricula in a few colleges and universities. Home economists promoted the application of research in areas such as sanitation and industrial efficiency to domestic settings and urged women to approach their new role as consumers with discrimination and education. For women, who were generally excluded from science careers, home economics provided a professional identity and culture. In addition, home economists claimed a key place among Progressive Era reformers who viewed more knowledgeable wives and mothers as central to a better society.[21]

Government support for home economics instruction created new career opportunities for women. The Smith-Lever Act, coupled with the 1917 Smith-Hughes Act's provision for high school training in agriculture and home economics, in turn, fostered demand for college curricula that would prepare people to fill these new positions. In the South, for example, George Peabody College for Teachers in Nashville led the way in training extension workers who would address rural problems. Soon after Peabody opened its doors in 1914, home economics was the first department to accept students. In the summer of 1915 its faculty offered courses in canning and club organization geared specifically to the needs of home demonstration agents. Home demonstration work and colleges such as Peabody formed a symbiotic relationship.[22]

As colleges offering home economics training proliferated, academic standards for agents rose. One development fed the other. By the 1920s, home economics training expanded to include courses in journalism, public speaking, rural sociology, economics, psychology, pedagogy, art, and literature, which would prepare agents to accomplish the myriad tasks that comprised their job.[23]

Professional standards increased as the responsibilities of agents proliferated and the scope of the home demonstration curriculum developed. The first agents focused on gardening, canning, and marketing. Their successors became domestic advisers, business consultants, community organizers, and social workers. Agents established home demonstration clubs, whose members studied lessons in health and nutrition, home management and interior design, textiles and clothing construction, landscaping, and marketing. Agents developed rural women's leadership skills, teaching them how to plan and conduct club meetings and encouraging them to participate in local politics and community affairs. Some home demonstration clubs, for example, provided hot lunches at country schools. As agents confronted the effects of the depression on rural women in the late 1920s, they also began to coordinate state and federal relief efforts. Besides giving demonstrations at club meetings, agents wrote domestic advice columns for county newspapers and delivered radio broadcasts.[24]

Harder to teach but no less important for an agent's success was her ability to deal with people from all walks of life. Good agents possessed the tact and diplomacy necessary to challenge how rural women cooked meals, decorated their homes, and cared for their children and to convince the women to try methods backed by scientific research. Even as they engaged in the cultural politics of housekeeping with ordinary farm women, agents kept apprised of county politics and solidified support for their work by socializing

Anne Pauline Smith, North Carolina district home demonstration agent, 1922–47, ca. 1915. (Photo courtesy of Franceine Perry Rees)

with the powerbrokers who paid their salaries. Political savvy thus proved as essential as knowledge of nutrition or dress design.[25]

Pauline Smith's career represents the transition to professionalism among home demonstration agents. As she became fond of saying, she had "grown up with" home demonstration work and cherished it "like a first child." From the outset, Smith excelled in her job. By 1920 the summary of her year's accomplishments won special praise from a superior, who scribbled across the front that it was among the best reports submitted. Smith displayed an ability to adapt to local circumstances, a knack for winning public recognition of her projects, and a gift for self-promotion. Because Franklin County farm girls had to work in family tobacco patches and lacked time to devote to individual tomato plots, Smith modified plans and encouraged them to can tomatoes grown in family gardens. A bread-making contest attracted more than four hundred competitors, and a newspaper account of the "bread campaign" included a long list of winners whose "white feathery" biscuits had brought "fame . . . to the makers." Smith reported that her most successful demonstrations included lessons on remodeling clothing, making hats, improving houses, and installing labor-saving devices.[26]

A no-nonsense administrator, Smith did not hide her light under a bushel. An advertisement for the Franklin County Fair boasted that Smith, the director of its women's department, was "not excelled by any Home Demonstration Agent in any state in the South" and concluded that she "has spared no efforts to make her department the best that will be seen in any fair this year." To achieve excellence, Smith established stricter rules that included barring children from the exhibit hall on Saturday and Monday because they would "be in the way of the directors."[27]

Smith was a go-getter, and in 1922 McKimmon promoted her to district supervisor. District agents were the workhorses and troubleshooters in the extension chain of command. They functioned as personnel managers who hired and assigned county agents. As supervisors, they evaluated local agents' job performance, advised them on how to improve their work, and investigated problems. Additionally, they became shrewd political operatives, who curried the favor of county commissioners and school board members who determined how much local governments would contribute to home agents' salaries and expenses. The district agents served as liaisons between local agents and superiors at state headquarters. District agents, one contemporary observed, were "the universal joints which connect up the different vital parts and make the machinery operate smoothly."[28]

Smith answered to the state agent, Jane McKimmon, but she enjoyed enormous autonomy. From her new office in the Beaufort County seat of "Little Washington"—where she boarded with her brother Clifford, his wife Tillie, and their young son—Smith oversaw a district that ran from the Outer Banks to the heart of eastern North Carolina's cotton and flue-cured tobacco country a hundred miles inland. Her duties required constant travel, as she consulted with agents, judged kitchen-improvement contests, spoke at home demonstration club meetings, and cajoled county commissioners. Once a month Smith joined fellow district agents in Raleigh for a conference with McKimmon. Otherwise, she was her own boss and set her own schedule.[29]

Smith assumed responsibility for a district in eastern North Carolina where backing for home demonstration was mixed at best. Although the state was considered a national leader in extension education, support ultimately rested with local officials. What counties could give, they could also take away. Sometimes local support shifted with prevailing political winds; at other times, it hinged upon the personality and performance of individual agents. In 1921, for example, continued endorsement of home demonstration in Pitt County depended upon the appointment of "the right woman" who "got results." In neighboring Martin County, "people who did not stand for progressive things" had elected new commissioners in 1920; their "first act" was to discontinue extension work. More specifically, their agent's poor performance had annoyed residents, who "did not feel as if they had gotten just returns for the money invested" in her position. In addition, tax revenues dwindled as farmers in eastern counties suffered from falling crop prices after World War I, prompting officials to keep a tight rein on public spending.[30]

During the 1920s, Smith devoted much of her time to supervising agents and politicking with local officials. A high turnover rate among employees combined with on-again, off-again funding kept her hopping. During 1924, for example, four different agents came and went in Onslow County alone. That same year commissioners in Hertford and Pasquotank Counties decided to let voters determine the fate of home demonstration work. "It required concentrated effort and the strong backing of the leaders of both counties" to halt the referendum movement, Smith reported. In Pasquotank, "two mass meetings were held before we could force the men to change."[31]

As Smith advanced on the extension career ladder in the 1920s, the professional standards for home agents were rising as well. A reputation for good housekeeping and civic-mindedness no longer sufficed as credentials. More and more of the new wave of agents entering home demonstration had earned college degrees. An administrator like Jane McKimmon could tap a growing pool of academically trained women and a cadre of experienced agents when filling positions. In 1925, for example, she boasted that of the twelve women hired that year, "nine of them have B.S. degrees from accredited colleges." The remainder matriculated at institutions that did not grant degrees, "but have had from five to ten years experience in Home Demonstration work." The extension service attracted women with degrees from such prestigious schools as Columbia University's Teachers College, the University of Chicago, and George Peabody College for Teachers in Nashville, as well as state and regional women's colleges.[32]

In addition to hiring new recruits with strong academic credentials, the extension service offered veteran white agents opportunities to continue their education in order to meet new academic expectations. They could take paid leaves of absence to complete courses or pursue college degrees. The middle-aged McKimmon, for example, plunged into studies at North Carolina State College and was awarded both a bachelor's and a master's degree by 1929.[33]

The new emphasis on credentials proved to be a mixed blessing. On the one hand, requiring more training could raise the professional status of home agents. On the other hand, higher standards could foster tensions and deepen generational rifts among agents, shutting doors for some women even as they opened them for others. In 1921, for example, a southern extension pioneer, O. B. Martin, praised early agents who, though lacking degrees, had nonetheless "set standards and established results which have become precedents and guiding stars for their successors," who enjoyed college educations. Martin worried that "new agents with more elaborate training are in constant danger of trying to teach too many things at once." To illustrate his point, he used the story of a state supervisor who admonished a young agent fresh from college to "forget" her formal home economics training. Although her "superior education" could be an asset, he concluded that a storehouse of experience gained from working with poor country-women would stand her in better stead.[34]

Pauline Smith greeted the higher professional standards with ambivalence. During the 1920s and 1930s she bolstered her academic credentials by taking summer school courses at a regional teachers college. Continuing her education presented Smith with exciting opportunities and at the same time added obligations to her already demanding schedule. Working and studying kept her burning the candle at both ends. She enrolled in courses, she explained, not "for what I would *learn* but for the units of credit." Although cynical about the value of coursework, she had to keep pace. "I cannot let little eighteen year old girls know more than I," Smith fretted. "The world is demanding better trained people & I cannot be at a standstill." By her calculations, the choice was either to "go forward or backward." Plainly, she felt the competition breathing down her neck and was threatened by a rising generation of women with more formal education.[35]

Despite increasing demands, home demonstration work offered women more career mobility than other middle-class occupations, notably teaching. Local agents occupied one end of an organizational chain that reached from county seats to state headquarters at land-grant colleges and ended at USDA headquarters in Washington. Joining a growing government bureaucracy had advantages. The elaboration of a hierarchy of specialists and administrators within the extension service created new professional opportunities. Subject-matter specialists, whose expertise centered on a topic such as foods, textiles, or home management, occupied a slot between local and district agents. The extension service also provided agents with a vast network of potential professional contacts.[36]

In addition to supervising the work in their purviews, state and federal administrators ran what amounted to employment clearinghouses for agents seeking new assignments in another county or even another state. Regional home economics field agents in Washington were key links in the process of connecting agents to job vacancies. In 1928, for example, the home agent in Currituck County, North Carolina, activated her extension contacts when she desired a change of scene. Rachel Everett, a graduate of the University of Chicago, worked as a home agent in Maryland before coming to North Carolina in 1924. Everett excelled in her job and earned kudos from her supervisors, but now she itched to move on. After reading that Hawaii offered a new frontier for home demonstration work, Everett wrote to the western states field agent, Madge J. Reese, inquiring about the possibility of a job there. Already acquainted with Everett's sterling performance, Reese sent her the name of the director for territorial extension work and suggested that she contact him. In the meantime, Reese offered to keep Everett in mind should vacancies appear in western states. Four months later, the Currituck agent wrote Reese again, asking for help finding a new position, preferably a promotion to district agent. Again, Reese supplied the names of state extension directors and described an opening for a home management specialist in Montana. Then in mid-June, Everett appealed to Florence E. Ward, home economics field agent for the eastern states, who promised to keep her in mind when vacancies occurred. As it turned out, Everett did leave Currituck, but she settled for a change that took her no further than Craven County, about a hundred miles to the south. Nonetheless, for agents bitten with wanderlust or seeking professional advancement, the extension network could help them satisfy the urge to move on or move up.[37]

A thickening bureaucracy carried vexations as well as advantages. Supervisors often undermined the independence and autonomy of individual employees. Most problematic was the increasing demand for annual reports and accountability. Even as the reports defined agents' identity as professionals who followed precise methods and recorded their accomplishments with care, compiling narrative and statistical reports became the bane of many agents' existence. Some labored for the better part of a month to complete the work. In December of 1929, for example, Pauline Smith complained that the "darn reports" from agents in her district had yet to arrive and she had not even begun her own. More than a month later she felt "oppressed" and as if she were "carrying some heavy load"; she counted "those reports" as the culprit.[38] But toting up successes mattered. Promotions and pay raises hung in the balance, and some years the very existence of the program itself might depend upon the achievements that agents could claim when county commissioners and state officials weighed whether to continue funding.

Throughout this process, agents created a professional culture in tension and collaboration with the clubwomen who were their constituents and the county officials who were their patrons. Regardless of how the extension service defined professional standards, agents served at the pleasure of county commissioners and club members, who judged their suitability for a particular community and evaluated their job performance with a more idiosyncratic yardstick. During a tour of home demonstration clubs, for example, Jane McKimmon once encountered a group who asked to confer with her privately. Their agent was efficient and well educated but, the women protested, "'she doesn't love us!'" "Efficient, yes, in a hard way," McKimmon commented, "but lacking love she lacked all as far as those women were concerned." Club members demanded empathy as well as expertise. Their complaints yielded results; to the women's delight, McKimmon assigned a new agent to the county. Club members could prove the final arbiters of agents' professional success.[39]

On clubwomen's scorecard, Pauline Smith won high marks. Sometimes she despaired at the intransigence of farm women slow to accept her advice, and she once cringed when a club member boasted that she had never read a book. But she reserved her harshest criticism for farm men and a rural male culture that sanctioned boorish attitudes toward women, and she bestowed her highest praise on women who used home demonstration to improve their lot and uncover hidden talents. Compared to their wives who cooked, washed, ironed, gardened, and worked in the fields, husbands appeared lazy and loutish. "If men tried as hard as these women," Smith observed, "how our land would blossom and prosper." After admiring a club leader who differed from her husband as much "as a diamond from glass," Smith concluded, "I love these people and the efforts they have put into their own personal development & into the community & home improvement. How they have grown." Clubwomen's respect for Smith was mutual.[40]

Some women thrived as home agents; others did not. Employee turnover was high, and in the mid-1920s McKimmon sought to stabilize her staff by setting a minimum age of twenty-seven for new hires. These women had made career commitments that pleased bosses, and they had acquired experience in life and work that stood them in good stead with their clients. Like one agent who realized "she was not fitted" for the extension service "because she was very slow in making friends," some turned to welfare work, taught school, or used home demonstration as a stepping stone to occupations in business. Others considered the job a way station between school and marriage and motherhood. Still others, like Pauline Smith, found it a satisfying career.[41]

FOR MORE THAN twenty years, Smith balanced a career and courtship. "This is really an infringement on rules," she confided to Frank O. Alford in

1929, "to attend a *millinery* class and write to one's sweetheart at the same time!!" While fourteen agents trimmed hats at a training session, Smith dashed off a note to her fiancé of three years. Even as Smith, known affectionately to beaus as "Polly," declared her love for Alford, she also professed devotion to her job. "No doubt you are tired of hearing of Extension work," she told Alford. "It is so much a part of my very existence, my very life, I probably bore people with it. I do not mean to do so. However, I feel that no work is worth talking about to a greater degree."[42] Efforts to reconcile the possibility of marriage with professional aspirations preoccupied Smith for the better part of two decades.

Seven years her junior, Alford was a native of Seven Paths in Franklin County who had recently completed dental school in Atlanta and was setting up a practice in Charlotte. His family farmed for a living, and Smith judged the male culture from which he had emerged as coarse, crude, and especially disrespectful of women. Over the years, Smith tried to polish Alford's rough edges, but he was doomed to wind up in the category of men who disappointed her. "You certainly do like to find fault," Frank Alford wrote in the summer of 1923. "You dislike everything on earth that I do and then fall in on my clothes."[43] Smith's insults directed at Alford's outdated checked suit were among the more benign barbs she would toss during the next twenty-five years.

The early 1930s proved to be years of personal and professional crises for Smith, who was now in her forties. Her engagement to Alford dragged on, lengthened in part because she believed he lacked sufficient ambition to build a lucrative dental practice and because she was reluctant to give up her career. In the meantime, Smith wrestled with her ambivalence about having children. On top of everything, support for home demonstration weakened as counties cut their budgets during the depression. Meeting challenges to the work that had provided her a good livelihood and formed such a central part of her identity left her at once exhausted and exhilarated.

As county and state officials parceled out declining tax revenues, public employees like Smith found their salaries reduced or their jobs threatened with elimination. In county after county, commissioners debated whether home demonstration work was expendable. Her entire district was "full of unrest," Smith reported to Alford in April of 1930; "I want to run away from it all." Yet lobbying politicians was the part of her job that she liked best. Her long letters to Alford described a wearing round of duties. She conferred frequently with local agents, sometimes delivered as many as four speeches a day, and devised strategies for maintaining support for agents' positions. "I have been *back in* the harness, pulling a heavy load this week," she wrote later in April. "I rather like the fight, if I am sure of winning." Challenges to funding in Beaufort, Pitt, Chowan, and other counties kept her moving to put out brush fires. "This has been the busiest week I *ever experienced*," she reflected in early May. "We are working to save Hertford."[44]

By the spring of 1932, the busiest weeks of 1930 must have looked like a calm before the storm. In letters written from hotels and boarding houses across eastern North Carolina, Smith confided her fears that extension work was "doomed" and outlined efforts to save it. The challenges inspired her. "I get all excited over the work," she wrote from Plymouth in March of 1932. "I am crazy over different counties. I love the contact with the politicians & leaders."[45]

One of the hardest-fought battles of the season occurred in Beaufort County, where the commissioners had voted to slash public spending and to suspend extension jobs altogether. When the commissioners arrived at the register of deeds' office for their July meeting, an overflow crowd greeted them, and the meeting had to be moved to an upstairs courtroom. At the top of the steps, the local newspaper reported, the commissioners "found awaiting them a crowded room, well over half filled with women, an unusual sight." Loyal home demonstration club members had gathered to protest the withdrawal of local agent Violet Alexander's salary.[46]

Clubwomen and Smith had orchestrated an impressive bit of political theater. A former congressman made a "surprise" appearance and delivered a "verbal barrage" in favor of extension work. Smith herself appraised the value of the home agent's services to the county. The 87,000 quarts of food that club members had canned, she calculated, were worth $17,000; their home sewing had augmented clothing budgets by $3,000; and women had sold over $25,000 worth of goods at the curb market. After Smith's speech, home demonstration members and 4-H Club girls testified to "the favorable points of the agent."[47]

Despite Smith's efforts to shape public policy, the women's testimonials fell on unsympathetic ears, and the commissioners struck the home agent's salary from the budget. A few days later, Smith conceded, "I have lost the hardest fought battle of my life," and the contest had left her "exhausted in mind, body, spirit." She fingered the culprits. "A few wealthy men did it all," she reported, "men who have no charity, no love for less fortunate people."[48]

The Beaufort County setback notwithstanding, Smith continued her campaigns. The first Saturday in August she headed to Hertford County, where, the week before, the question of whether to renew the salary of the agent had resulted in a tie vote among commissioners. Smith joined local home demonstration advocates who hoped to sway at least one vote in their favor before final budget approval. She lobbied businessmen and politicians until after midnight, pleading with one recalcitrant commissioner "as [she] had never plead [sic]" before and finally chalking up a victory. On Sunday she traveled to another county in her district and roused a commissioner from bed at 11:30 P.M. "He came out . . . nearly nude," she told Frank, "but I was there for business and his help. We forgot clothes as I talked to him."[49]

As a season of hard-won victories and disappointing defeats drew to a close, Smith spent a rare day off with her brother's family in "Little Washington." "I know this work was part of my very being," Smith reflected in a contemplative letter to Alford, "but I did not know how nearly a part of me it is until we have begun to lose counties." After helping her sister-in-law with domestic chores and contemplating a choice between keeping house and working as an agent, Smith concluded, "I had rather interview politicians *and fight* than plan meals."[50]

And fight she did. In Beaufort County, private donations covered agent Violet Alexander's salary until Smith could organize a letter-writing drive on her behalf in the winter of 1933. She burned the midnight oil as she arranged transportation so that fifty clubwomen could gather and bombard commissioners with letters and telephone calls appealing for the restoration of home demonstration funds. The clubwomen "put it squarely up to the men that the women are voters and they expect to exercise their right." The hardball tactics worked. Although the commissioners failed to provide the full amount requested, they did reverse the past summer's decision to rescind support altogether.[51]

The battles that Smith undertook in the name of local agents were waged on her own behalf as well. Just as counties reduced their contributions, so the state reduced its share of the extension service budget. Along with other public employees, Smith suffered a 20 percent salary cut and lost most of her expense account. Rumors circulated that extension staff reductions were imminent, and Smith feared that she might be among the first to go. "We are all worried, I know this," she admitted. "Worried for our jobs and for our agents."[52]

Staff reorganizations taken as economy measures added extra burdens to district agents' workload and eroded their prized autonomy. Smith and her colleagues faced a white-collar speed-up and stretch-out. After district offices closed in the fall of 1930, supervisory agents worked in Raleigh half of each month and spent the remainder traveling in their districts. Smith had more territory to cover with three additional counties in her region, and she enjoyed less clerical help now that she shared a stenographer with coworkers. Perhaps most galling of all, she was subject to the scrutiny of superiors and slave to an office routine in a way that she had never been before. Smith chafed at the new circumstances. "I am happy out in the field," she wrote Alford in 1932. "I realize you have heard this a million times. But at home I could rest if I was sick & make it up. Here you have to be at the office at 9 o'clock & not leave before five. You see I am just spoiled. I really would like to cry over it. We are all tired." Smith decided to take a cue from a fellow district agent who maintained her own schedule even at state headquarters. "She stayed in bed this morning," Smith noted, and arrived at the office at eleven. "She is a fine worker but when she wants to she rests—a good policy."[53]

Being constantly on the move and "living in suit cases" took their toll. During a spring 1933 tour of coastal counties, for example, the bus that Smith was riding to Manteo broke down three times. While a mechanic repaired the bus, passengers wilted in the heat and swatted mosquitoes so voracious "that we had to sit on our feet to save our legs," she told Alford. "I was glad to crawl out at five o'clock in the morning and go to Currituck to work." Despite the demands of the road that often left her feeling "dead," Smith was "ready to go again" after she had spent a few days in Raleigh. "I am not happy in this office," she complained, "though I should be. Only occasionally are we 'jacked up.'"[54]

The uncertainty that surrounded Smith's engagement to Alford only compounded her worries. But the tensions that smoldered and flared helped Smith clarify what her work meant to her and to define her expectations of marriage. The couple conducted a long-distance romance for more than two decades. The protracted affair stemmed in part from Smith's judgment that Alford was not ambitious enough and in part from her refusal to surrender economic independence and a lively professional life. While Smith wanted to hitch her wagon to a rising star, in her eyes Alford sputtered, stalled, and refused to soar.

Four years into their engagement, Smith pressed Alford to determine if he could "*support* a wife in 1931." She faced middle age and stood at a turning point in her career with the extension service; she needed to make some decisions. She wanted to continue taking college courses to bolster her professional credentials. At the same time, she contemplated having children. The path she would follow hinged upon how their relationship unfolded. In the spring of 1930, Smith lamented the passing of her youth and the time that she had invested in work that sapped her "pep." "My heart just aches," she wrote, "when I think of the vain years—helping *the public*. Why, now is the time I should have my home & the public work could come in late middle age. Oh, time in thy flight!" When she encountered children, her ambivalence about motherhood surfaced. "Some of these times," she told Alford, "if I ever get so I can *stand them*, I am going to adopt two children, if you will *let me*." Yet later in 1930 Smith declared that she shared "the same desires as other women. I have wanted a home, children, and all the things that an ambitious woman wants." Her age was reducing the chances of a safe pregnancy, or pregnancy at all. "It nearly breaks my heart," she concluded. "I did want children to carry on, in the ways that I have failed. It isn't my fault that fate is handing me this deck."[55]

Smith shuffled and reshuffled the cards of her life. But in the end she decided that despite the toll her job exacted and all the uncertainties surrounding it, her career was her ace in the hole. In fact, part of Smith's professional identity was forged in response to Alford's aspersions. His offending letter does not survive, but in 1930 he made what his fiancée deemed "sarcastic" remarks about her job. "This work has made me!" Smith shot back. "But for it, I would be a little county teacher. It has helped others and helped my family. . . . I can remember when I did not have a penny. *It has made me* independent." Her "greatest desire" was "to be so independent, so well established, educationally, financially & socially, that I can tell all to go to ――― if I so desire." Among the benefits of her work, she continued, were "*social recognition, power* to *make* people recognize me," and "an opportunity to be of service to people. I do love it & shall always do so. It has taken the place of social life, of home, friends, & recreation. It has been my life. It is part of me, from which I could not be divorced."[56]

Although a growing proportion of American women combined marriage and career, Alford apparently expected Smith to stop working when they wed. She, on the other hand, could not imagine trading a job for the "grind" of housekeeping and exchanging a salary that she could spend as she pleased for a "modest allowance as a wife."[57] To ask a woman to abandon her career for "cooking, planning meals and the monotony of housekeeping or sitting around," Smith reasoned, "would be like caging an eagle. It can't be done."[58] She invoked the prerogative of age and put the younger Alford in his place when she asked in 1932, "Child, do you think that I will change if I would marry you?" No, she answered. She would retain her "opinion" and "desires," her own "wish to lead, to be in action, to be *looked to*, to be considered outstanding." Why should she jettison her work? "A man does not give up a career for a woman," she pointed out. "She would not fill his life. A man will not fill mine."[59] She might have to tolerate petty office politics, but Smith enjoyed interesting colleagues. "You would have to do a mighty lot," she informed Alford, "to take the place of the people here and my independence and salary."[60] Her career, moreover, safeguarded her against insult and abuse. When Alford once reprimanded her for spending the money she earned to buy flowers, Smith "decided then and there that I would never be so I could not make my own money." "I want to make my own money so long as I do live," she declared. "I had rather die than be dependent."[61] After he once spoke harshly to her, she knew why she treasured her job. "Dependence is one of the great dreads of my life," Smith wrote. Should he use a caustic tone or be unkind again, she advised, "I would be able to pack my things and walk out. . . . I just will not stand for it."[62]

Although Smith sprinkled her letters to Alford with terms of endearment, her tongue and pen dripped their share of acid. Even as she described kissing his picture each morning and night and longing for him, her letters often inventoried his shortcomings and suggested ways to overcome them. Her advice ranged from Benjamin Franklin–style maxims to blistering excoriations. As Smith's *"high ambitions"* and desire "to be a *real inspiration"* met Alford's inertia, the relationship soured but endured.[63]

To Smith's way of thinking, many of Alford's faults could be traced to his upbringing in a family in which rural men devalued and disrespected the work of women. By the time of their engagement, Alford's father was dead, renters tended the family farm, and his mother boarded in rotation among relatives. Smith contrasted her own early home life with his. While her father had risen early to start the fires for her mother, Alford and his brothers had slept while their mother began her day's work. While her mother had refused to accept lodgers in order to protect her daughter from association with strangers, Alford's family had taken in "illiterate farm hands on a plane of equality," thus diluting any hopes for "high standards, high ideals and accomplishments." One unfortunate consequence was that Alford's sister, Maggie, had received little encouragement and developed little desire for an education. Now she boarded in Raleigh and cut and curled women's hair for a living. Seven Paths, in Smith's estimation, was home to rubes and ne'er-do-wells who turned their backs on education, thrift, and pride. "All the lack of little courtesies which I get after you about," Smith scolded, "result from that early environment which might have been corrected in childhood. . . . Polite people do not dip soup bowls, pick teeth in public, [or] use knives improperly."[64] One could take the man out of the country, but one could not take the country out of the man. When "I see you yawn and stretch where there are women," Smith observed, "when you sit as you do at times in public places, I know that habits formed among [your family] are deeply laid; when you talk as you did Sunday about woman's work, I know that you have a far piece to go yet to get away from traditions, customs and inhibitions made with that set."[65]

As a fiancée, Smith adopted methods of instruction that she used as a home demonstration agent trying to change the habits of rural women. After noticing the dark circles under Alford's eyes, "the sallow skin—the flabby muscles," she instructed him to walk, to swim, to eat more vegetables, to drink more water.[66] Drawing upon her experiences instructing women how to accomplish their work more efficiently, she devised a detailed daily schedule for Alford that allotted time for work and recreation.[67] When he balked at her advice to sleep in a cool room, he reminded her of "some of the country women with whom we work sometimes—we can . . . give them the science for certain practices—and they continue the old way, even though that way, brings on T.B., hardened arteries, high blood pressure, pellagra. It is hard to get people out of childhood practices."[68] Smith justified offering Alford candid advice and making bold assessments of his character because she had years of experience sizing up agents under her supervision.[69]

Although the depression certainly hindered Alford's ability to build a lucrative dental practice, Smith placed the burden of responsibility squarely on his shoulders. In fact, meeting adversity head-on was a sign of manhood. "Men with many dependents are fighting this out & fighting it out bravely," she wrote. "Don't be a weakling! Be a man!"[70] Rather than blaming outside circumstances for his misfortunes, he should join a church and civic organizations in order to build social and professional networks. "God, Frank, if you could see where your apathy . . . is leading you and me!" she stormed in 1932. "It is not sissy to do these things. Every big leaguer in the country, all good literary men know the Bible. . . . You give me your promises & never carry them out."[71] Again and again, he provoked her ire. "One of the traits which I despise about you," she later seethed, "is never being able to make a decision. . . . I hate indecision!"[72]

The fractious couple continued their engagement as personal and professional demands mounted for Smith during the latter 1930s. In 1936 Smith's brother Clifford disgraced the family when he was caught mishandling Works

Progress Administration funds, and she assumed financial responsibility for his wife and son.[73] In addition to family worries, Smith maintained a frantic professional pace. She juggled new responsibilities as acting extension home beautification specialist with the familiar duties of district agent and college courses in Raleigh. By the summer of 1936, the stress of travel, family turmoil, and studying brought Smith to the breaking point. One June morning she cried during her first class and "just boohooed" in the second.[74] Although Alford insisted that he was in a financial position to marry in the fall of 1936, Smith suspended the engagement, claiming that "duty" to her family came first and that she would only be a burden. "For your future peace and happiness," she wrote Alford, "you do not need me. I am old before my time—I am a wreck." In December of 1937, she once again released him from marriage commitments, "the best Christmas gift a woman ever gave a man in your position."[75]

How Alford responded to Smith's complaints is unclear because only a handful of his letters survive. "You will never know how much I love you," he insisted in early 1938. "I have missed you so much since you came up here and since Christmas."[76] After moving into a new apartment in Charlotte, he wished that they were married. "I have definitely decided that I need a wife as soon as possible," Alford asserted in July. "No place can be a home," he believed, "without the touch a woman can give it."[77] But nine years would pass before his place received the transforming woman's touch.

If Smith constantly put Alford into a no-win situation, perhaps it was because she was boxed into one herself. A Dorothy Dix advice column that Smith clipped in 1930 offers a sobering reminder that it was easier in theory for a woman of her generation to combine marriage and a career than it was in fact. After a correspondent asked Dix to describe "the greatest bar to women's success in business and the professions," she answered simply, "Sex. Being women." And there was nothing that they could do about it. While a boy grew up expecting to work and looking "forward to achieving fame and fortune at the end of a lifetime of faithful service to his occupation," a girl knew "her real job in the world [was] being a wife and mother" and considered any other work "a mere makeshift." Employers refused to invest time and money to train women because they believed matrimony would soon lure the women away. When a man married, he "simply annexes the comforts of a home and the joy of having wife and children to his occupation." A woman enjoyed no such luxury and had to choose between love and her career. "If she tries to keep on with her work," Dix observed, "it imposes an intolerable burden upon her because she has to do the work of both a man and a woman." Should she try to do both, her husband would grouse because he did not have her undivided attention; if she had children, she would be worried about how well "some hireling" cared for them while she worked.[78]

Smith shared the dilemmas of contemporary professional women who struggled with "the contradictory imperatives" of domesticity and career.[79] During the first decades of the twentieth century, new professional opportunities held out the promise of economic independence. At the same time, women revised their expectations of their male partners. Modern ideals of heterosexual relationships stressed mutual trust and emotional intimacy. But rising personal and professional expectations often collided rather than meshed. Pauline Smith, for example, wanted a partner who both spoke to her with a "kiss in the voice" and earned more than she did. "A $4000 woman," she observed, "would never be happy with a $2000 man. I have seen it tried too many times."[80] The idea of combining marriage and career created more contradictions than either Smith or Alford could reconcile.

Smith's job as a home demonstration agent generated other dilemmas as well. The conditions of her work made it impossible for her to practice the very gospel that she preached. While she encouraged rural women to create artfully landscaped homes appointed with tasteful pictures and furnishings, she lived in rented rooms. While she taught women to prepare balanced diets, she skipped meals or ate restaurant food that gave her indigestion. While she advised women on how to accomplish their work in ways that would save energy and health, she complained constantly of fatigue and illness. In 1939, Smith sympathized with a colleague who was looking for another job because "she cannot keep up with it all" and had "no time to live, no fun & life [was] fast passing away."[81]

By the 1930s, Pauline Smith found herself in something of a career box as well. She felt exploited when she had to carry the "double duty" of specialist and district agent for five years. "The result is impaired health and a terrible disposition," she confided to a colleague in 1938. "Perhaps too late I have learned my lesson but you just watch and see if ever again I do what I have done in the past."[82]

To resolve some of these personal and career paradoxes, Pauline Smith created a role that one agent under her supervision dubbed "Professional Mother." In the company of colleagues, Smith enjoyed some of the emotional satisfaction that Alford denied her. Nurturing the agents under her supervision was like raising a professional family.

A demanding boss, Smith fostered a tremendous loyalty among the women who worked in her district. One way in which she cultivated a sense of group identity was through a club known as "the Huddlers." Home agents in the northeastern district chartered the club in the winter of 1931, when local funding fights raged, and they continued to meet for a decade. Although descriptions of their meetings grow thin after 1934, the minutes that do survive suggest that the women used the group to escape the pressures of work, to build staff morale, and to mark personal rites of passage. At their first meeting in Norfolk, Virginia, the Honorable Order of the Huddlers adopted the buoyant balloon as their emblem, claimed as their theme song "We're Here for Fun Right from the Start," and decided "that our pass sign shall be a smile." They took in a show, shared a suite at the Monticello Hotel, investigated a cabaret, savored lobster, and shopped. Over the next three years, the Huddlers rented a cottage at Nags Head and enjoyed icy spring dips in the Atlantic; they met at a Beaufort County hunting lodge that they later discovered was reputed to be a road house; they flirted with attractive widowers while boating on Bull Bay near Columbia. In March of 1932, they feted an engaged Huddler, Edna Evans—who had "postponed her last date before her wedding to attend the meeting"—with an original advice booklet entitled "Sense and Nonsense." The next month they read a thank-you note from Evans, now described as "'evaporated huddler' Edna Evans Bells." During the final recorded gathering in the winter of 1941, the secretary noted, "We have two unwritten laws: Shall not talk 'shop' at huddle" and "Must not bring anyone that would cramp our style."[83]

The Huddlers solidified the friendships among agents that had long been central to their professional culture. Agents worked together and played together; they shared holidays and marked special occasions with the exchange of gifts. In 1927, for example, Smith gave Jane McKimmon a bag of lavender for Christmas. In her letter of thanks, McKimmon noted that she got "a delicious whiff every morning when I take out my fresh teddies."[84] In 1932, single and widowed agents living apart from relatives in Raleigh celebrated Thanksgiving by going to a movie together and eating until they groaned.[85] And on at least one occasion Smith and McKimmon slipped away from the office early enough to take in a matinee.[86]

Although the Huddlers vowed not to spoil their fun with "shop talk," the friendships that Smith cultivated among her district's agents influenced her supervisory style. In 1939 when she heard the buzz that an agent identified only as Mary was dating a married man named Mr. Knight, Smith asked another home agent to investigate. The intermediary reported that Knight was divorced and doubted that Mary had been responsible for his estrangement from his wife. The man worked for a Norfolk nursery and had met Mary when he delivered an order of shrubs for her clubwomen. The romance blossomed, and the couple had courted quite openly. "I cannot believe Mary has intentionally hurt anyone or knowingly taken a wife's husband from her," commented the colleague-turned-detective; "neither do I believe she has met him places where it was not perfectly all right to meet him."[87] A few days later, however, Smith traveled to Norfolk to interrogate Knight's employer and to check out the gossip herself. "It is true!" Smith discovered. "It is my unpleasant job to tell her Monday that her fiancé is a married man."[88] Whatever the precise circumstances of the romance, the couple eloped a year and a half later, and Mary T. Knight sent Smith a 1941 New Year's greeting from Florida, where they were honeymooning. "I didn't want to spoil Christmas for you," the agent wrote, "so I waited to break the news."[89]

On another occasion, Smith consulted a friend about the job performance and private life of the home agent in Gates County before she made personnel decisions. The informant described the love life of Ona Patterson, who had broken one long-term engagement and was now dating a school principal. "He gave her an electric blanket for Christmas," the correspondent wrote. "I wonder sometimes if Miss Patterson will ever marry, I wonder if she will ever love a man as much as she does her work." Patterson was an agent who did not "let her personal affairs or engagements" come ahead of her job.[90]

Agents who did choose marriage or motherhood often reported their decisions to Smith before almost anyone else. They approached her with some trepidation, but eagerly sought her approval. "Miss Smith," Eloise B. Perry wrote in 1941, "I have something to tell you which I know will disappoint you. And believe me, I would have had it different if I could have." Smith was the first to know that Perry and her husband had a baby on the way. "I haven't even told my mother," the agent confided.[91] In 1944 agent Anne Harris requested a two-week leave so she and her soldier-fiancé could marry and take a honeymoon. "My marrying will not affect my job at all," Harris assured Smith, "—unless it makes me a better Home Agent—for I know I'll be a happier one." The only people to hear the wedding news before Smith were Harris's parents. "I wanted you to be one of the first to know," the agent explained, "because you mean and have meant so much to me. You've indeed been a 'Professional Mother' to me in the past 2-½ years. I have their blessings, now all I need to make my happiness complete is yours." "You have my blessings, my love, and my best wishes," Smith replied. Smith speculated that Harris might later regret that she had cut her "girlhood" short, but she concluded, "I think every girl should marry, after she has prepared herself for any emergency, and you are prepared and have shown that you are equal to any situation that may arise in the future."[92] The degree of Smith's involvement in the personal lives of her agents became evident when Iberria R. Tunnell expressed regret that her boss had missed her wedding. The newlywed wrote, "I am anxious for you to meet 'our husband.'"[93]

Home demonstration agents of the 1940s talked optimistically of combining careers and marriage. Nonetheless, Smith must have smiled thinly in 1946 when a male colleague from Georgia shared with her a joke about ungrateful husbands that she might want to tell "in some of your meetings where some prospective brides are present who are about to swap a good $200 a month job for an opportunity to bathe skillets on a $15 a week allowance."[94] Working as a home demonstration agent had allowed Smith to escape such a compromise.

When Pauline Smith began her extension service career as a Tomato Club organizer in 1913, service to others had dominated home agents' public rhetoric and, indeed, was a primary motivation for their work. But for Smith, and no doubt for countless others, work as a home demonstration agent sanctioned personal ambitions, satisfied a hunger for independence, and provided the camaraderie of a female professional world. In a society where marriage and motherhood still determined women's destinies, work as a home agent offered some women the possibility of escaping or delaying those fates by making a career of teaching rural women to be better wives and mothers. Despite the apparent contradictions, Smith and other agents viewed their work as providing similar opportunities for the rural women who joined home demonstration clubs and adopted the advice they offered. Agents considered themselves allies and advocates for women who were trying to create more equitable gender relations on the farm and in rural life. Rural women could use lessons learned from agents to improve their health, to prod husbands to invest in modest household conveniences that made their work less burdensome, to appreciate their unpaid domestic labor, and to earn money that enhanced their independence. In much the same way that the Honorable Order of the Huddlers sustained agents, home demonstration clubs created female spaces where rural women could affirm the value of their work, continue their educations, develop their talents, and shape their communities.

When Pauline Smith retired in 1949 to concentrate on a marriage that proved as contentious as the engagement that had preceded it, she surveyed her achievements with pride. Support for home demonstration rested on a firm footing in her district, thanks to her vigilance.[95] But as she had helped to raise successive generations of "daughters of the New South," she may well have wished them an easier time of resolving the personal and professional paradoxes in whose web she had struggled.

CHAPTER FIVE

Women in the Middle

IN MAY 1922, Emma L. McDougald became Wayne County, North Carolina's first black home demonstration agent working for the state extension service. To recruit women interested in learning better ways of growing and canning food, she advertised her programs at church services on Sundays. Frustrated by the tepid response, McDougald then "went out in the fields and everywhere that I could get a woman or family to listen to me." By December she had convinced 235 women to join fourteen home demonstration clubs.[1]

During the next decade, McDougald worked against long odds to bring the benefits of home demonstration to African American farm women. Most of her constituents were poor—only 17 percent of black Wayne County farmers owned land—and she herself enjoyed few of the resources available to her white colleagues. "The first question that I ask myself when I go to a community," McDougald explained, is "what is the greatest need of the people and how shall I attack it?" To take stock of their problems, she spent the night with families, surveyed their farms, and visited schools and churches. "In nearly every case," she concluded, "I find the greatest need is to make the best of what they have." McDougald made the best of what she had, too. Unlike white home agents whose salaries or travel budgets allowed them to travel around the countryside in cars, McDougald relied upon trains or her own two feet to reach demonstrations. The train schedules forced her to arise long before dawn and return home to Goldsboro long after dark. "I am tired when I get to work some days," she observed, "yet, I work all day and part of the night."[2]

Despite the obstacles she faced as a black woman in the Jim Crow South, McDougald navigated a risky course to become an effective community organizer and advocate for the women she served. Conducting a home improvement contest in 1927, for instance, she found that accomplishing even modest changes in the kitchens of black women required delicate negotiations with the white landlords who owned the farms the women and their families tended and the houses that they called home. But her efforts paid off. One landlord built two new kitchens and ceiled two old ones in houses on his place; another "did a great deal to make a kitchen comfortable" for a contestant. Because all thirty-two participants had tried to improve their quarters, McDougald declared that all were winners. Although she dreaded asking local merchants to donate contest prizes, she was encouraged when "our white friends . . . responded with that beautiful spirit which helps more than value received." She was even more satisfied when black allies of extension contributed most of the awards. "We are glad of this," McDougald commented, "because the help which does one most good is self help."[3]

Black home agents like Emma McDougald were women in the middle. On the one hand, they answered to white officials who kept tight fists on the state resources parceled out to African Americans. On the other, they answered to members of black communities who sometimes doubted that any representative of a racist government—even a black woman—had their best interests at heart. As McDougald's language and actions suggest, she built on traditions of self-help and mutual aid and a philosophy of uplift among African Americans to justify the expansion of state services to rural North Carolinians whose access to power was severely limited in the age of segregation and disfranchisement. Her language invoked black pride, yet sounded safe to white patrons. McDougald praised "white friends" even as she criticized the handicaps under which she labored. She goaded white officials to share more resources with her and her constituents and described achievements confirming that rural blacks merited government services.

Until the mid-1930s, North Carolina's extension service sponsored only six or seven black home demonstration agents. To ignore the handicaps under which they operated—low salaries, inadequately equipped and understaffed offices, and burdensome workloads, to name only a few—would be naive and disingenuous. But to recite the disabilities alone obscures what agents did accomplish despite all the obstacles that discrimination placed in their paths and the contradictory consequences that public policies had for black farm families and the professionals hired to serve them.[4]

FROM THE START, Seaman A. Knapp assumed black farmers would be included in his agricultural demonstration program in the southern states. As early as 1906, Knapp drew on established African American networks at Tuskegee Institute in Alabama and Hampton Institute in Virginia, the leading proponents of industrial education among the region's blacks. Both schools already practiced extension work of their own design, using community outreach and improvement as part of their regular lessons. Beginning in the late nineteenth century, Hampton and Tuskegee sponsored annual conferences that attracted hundreds of rural families to hear talks on better farms and better homes. When Knapp could not convince agriculture officials in the South to underwrite the work of black agents, he used grants from the Rockefeller-funded General Education Board (GEB) to hire a few black men. Thomas M. Campbell of Tuskegee became the first black agent in 1906, and he eventually supervised black farm agents in the Deep South. John B. Pierce of Hampton was Campbell's counterpart in the upper South.[5]

When Congress adopted the Smith-Lever Act in 1914, key provisions ensured that black agents would receive the short end of the stick. Heated debates swirled around the question of how the extension budget would be administered in states with two land-grant colleges, one for whites and one for blacks. Southern lawmakers proposed that whites, who would use their discretion when disbursing money for extension work among African Americans, would control all funds. Senators from several northern and western states and members of the National Association for the Advancement of Colored People (NAACP) protested that black farmers would receive few benefits if white campuses controlled the funds. In the end, the arguments of southern senators prevailed, and the Smith-Lever Act allowed state legislators to decide which college would manage extension work. Of course, they all chose the white schools. The bill, moreover, prohibited future GEB contributions, thus eliminating a vital source of support for African American agents. After Congress passed Smith-Lever, the NAACP appealed unsuccessfully to President Woodrow Wilson to veto the legislation "on the grounds that it discriminates against the Negro farmers of the South." The money, the NAACP predicted correctly, would not be distributed fairly.[6]

To deter threats to funding or challenges to white colleges' control, federal officials warned southern extension directors not to ignore black farmers altogether. Thus, despite predictable opposition, southern directors retained a few black agents and advised white agents to devote some time to black farmers.[7] In the winter of 1915, for example, W. B. Mercier of the U.S. Department of Agriculture (USDA) advised C. R. Hudson, North Carolina's chief agricultural extension agent, to pay particular attention to black farm owners. "Personally," Mercier confided to Hudson, "I have always felt that the Government should help the negro farmers to some extent. We have them and are going to continue to have them," and, he maintained, whites benefited when blacks became better farmers. Hudson assured Mercier that several white agents worked with black farmers and that he planned to hire one or two more men to join the seven black agents already on his staff. Mercier agreed that the state employed enough black agents, but he suggested "there should be a little more thorough and systematic supervision of their work." Ultimately, Hudson decided that white district administrators would oversee black agents, but the state extension service might also send "one or two of the more progressive Negro agents on a tour of inspection" among their peers. White district directors, Hudson observed, "do not object to performing this supervision by giving advice personally to Negro Agents, but they would not want to ride around with them very much on warm summer days." In the face of such ugly prejudice, federal agriculture officials concerned about criticism of Smith-Lever's racial politics prodded southern extension directors to pay token attention to blacks.[8]

While North Carolina agriculture officials hammered out policies regarding black farm agents, home demonstration among African American women and girls proceeded under the auspices of public school officials. Not long after Nathan Carter Newbold, a white educator, became the state's first agent of Negro Rural Schools in 1913, O. B. Martin of the USDA offered to support the summer salaries of teachers who would organize "gardening, canning and crop work among negro boys and girls" similar to the clubs sponsored for white children. Martin also anticipated that the teachers would work with adults, "especially with the idea of improving home and living conditions."[9] The backing that Martin offered came from the GEB. Even after the Smith-Lever Act forbade the philanthropy's subsidies, Newbold circumvented the extension bureaucracy and channeled support for home demonstration through the public education system.

Black teachers brought to their summer work with Homemakers' Clubs finely honed skills as community organizers. Many of the women oversaw industrial education in rural schools, their salaries paid by the Negro Rural School Fund, a philanthropy administered by the GEB. This support continued the mission of the Anna T. Jeanes Fund, started in 1909 by a Quaker heiress to support African American education. Jeanes teachers consulted with instructors in schools all over the county under their supervision, and they sought the advice and support of ministers and parents. They lobbied white officials on behalf of black teachers and students and stretched slender financial support from local boards of education. Jeanes teachers who were graduates of Tuskegee, Hampton, and black normal schools often modeled their work after the community outreach projects inaugurated by their alma maters. Under the guidance of Jeanes supervisors, black schools came to resemble settlement houses where parents as well as children gathered to learn about health, sanitation, and nutrition; Jeanes teachers defined education expansively to include "the welfare of the whole community, and of the school as an agency to help people live better."[10]

Annie Welthy Holland's work exemplified the broad vision of Jeanes teachers. By the time she became the industrial education supervisor in Gates County in northeastern North Carolina in 1911, she had attended Hampton Institute and taught for nearly two decades in southside Virginia. Always aware that education continued beyond the schoolhouse door, Holland organized a cooperative to encourage land ownership among farmers and gathered donated garments to help clothe her poor students. When she began work as a Jeanes supervisor in North Carolina, Holland believed that schools, homes, and entire communities should work together to educate African Americans.[11]

In 1914, Holland eagerly used the newly available funds for home demonstration work to organize Homemakers' Clubs throughout Gates County. Like so many of the agents who would come in her wake, she sought the support of the church. While her primary duty was to help women and girls grow gardens and preserve vegetables, conducting demonstrations in homes gave her the "opportunity of teaching more than mere canning." Instruction on diet, health, and sanitation flowed naturally from conversations around the canning pot. Parents as well as children, fathers as well as mothers, took an interest in the work and produced tangible results. Although late summer rains followed by scalding sun reduced the yield of gardens, Holland's club members set a state record by putting up three thousand jars of vegetables. Among all of her duties, Holland considered organizing Homemakers' Clubs "*the best*" part of her work. "I feel indebted to you and other friends of your race," Holland wrote Newbold, "for enabling me to be of some service to my people."[12]

Holland's initiative and diplomacy as an intermediary between black constituents and white patrons opened a wider field of service. In 1915 she became the Colored Supervisor of Home Demonstration Work, overseeing forty-four county Jeanes teachers who spent their summers organizing clubs and helping members garden and can. For salaries, the GEB chipped in one hundred dollars and local school superintendents added another twenty-five. Homemakers' Clubs proved to be popular among both students and teachers, who sometimes contributed money from their own pockets to purchase canning equipment because they recognized the value of their demonstration projects. The preserved fruits and vegetables enriched lean winter diets, and girls and women sometimes earned enough by selling canned goods to buy items for themselves as well as to help purchase school supplies.[13]

In 1917, the United States' entry into World War I sparked increased interest in food production and prompted an expansion of home demonstration among black as well as white women. Home demonstration agents worked closely with the North Carolina Council of Defense, whose members organized the home front and whipped up patriotism—and in some instances policed civilian loyalty to the war. Jane Simpson McKimmon, state home demonstration agent, reported that white agents answered the "eager calls from the colored people" who wanted to contribute to the effort by learning how to can more food. "Intelligent colored women and men," she commented, "have come out to the [canning] demonstrations and agreed to go back and instruct others of their race in the art."[14] In 1918, McKimmon received federal money to hire emergency agents for the summer, and she expanded her force to include nineteen black women; the next year the number rose to forty-one. Most of the women hired were experienced club organizers recommended by Annie Welthy Holland.[15]

From one perspective, white and black home agents who shared common goals represented a fragile new frontier in interracial cooperation. White agents gained important insights into African American life when they crossed the color line to address church congregations and instruct black women. They developed respect for hardworking black colleagues like Dazelle B. Foster, who had "made a splendid impression upon the colored people" and whose "energy, tact and diplomacy" had impressed the white people of Davidson County.[16] Interracial collaborations symbolized the best of a reformist spirit fostered by World War I.

But a wartime culture that nurtured promising new possibilities in race relations also sanctioned coercion of labor and judged "loyalty" and "patriotism" as paramount virtues. Within this context, offering home demonstration to African American women assumed ambiguous and contradictory meanings for white and black alike. No doubt, expanding extension services to African Americans was part of a home-front strategy to stabilize a rural workforce that the military draft and wartime factory jobs had disrupted. White landowners in eastern North Carolina grumbled about a shortage of labor—or a scarcity of workers willing to chop cotton or prime tobacco for prewar wages—and refused to pay what they considered the "outrageous" sum of $1.25 that field hands now demanded for a day's work. White women, too, complained that they faced Mondays and Tuesdays without the usual complement of black women to wash and iron for them. "With the shortage of help almost universal," McKimmon observed in 1919, "the housewife has been eager to learn modern methods of laundry work," and white home agents invited manufacturers to explain the use of washing machines. Federal extension officials believed that black agents stemmed the exodus of workers by helping black farm families to achieve prosperity and satisfaction in the country.[17] Oddly enough, the most powerful advocates for increasing home demonstration's services to more African Americans may have been just those rural blacks who abandoned fields, kitchens, and wash pots for better-paying jobs elsewhere. Hiring a few home demonstration agents in hopes of retaining field hands and domestic workers might have seemed like a small price to pay.

Amidst concern about the availability of cheap farm labor, southern African Americans found themselves subject to the scrutiny of state defense councils, and agricultural extension agents were part of the government apparatus that monitored the labor supply. In the South, "work or fight" laws caught "idle" black women as well as men in their dragnet, branding them as unpatriotic loafers, and sent them to jail on trumped-up charges of vagrancy.[18] White men assumed that black women with young children should work for wages while their white counterparts tended to their families. One white farm agent in North Carolina, for example, let USDA officials know that he had his eye on black women workers. C. H. Stanton thought the "hundreds of Negro girls and women" in Louisburg could better spend their time as field hands picking cotton than as nurses caring for white babies. "These girls and women," Stanton wrote, "could quickly acquire the art that they seem to almost inherit, and would replace many a man that is busy elsewhere." Bradford Knapp—who took over at the USDA after his father, Seaman's, death in 1911—told Stanton that he could "see no harm in agitating the matter." But the black women who really galled Knapp were the thousands "who are doing nothing because of the extraordinary wages now paid husbands and fathers" serving in the army or working in factories. "If the number of negro women and girls who are idle today because of soldier's pay and the high wages paid by industries to negro men could be put to work I believe there would be plenty of labor without using the negro nurse girls."[19] Black women, then, found themselves subject to surveillance, slander, and intimidation during the war.[20]

To answer whites who charged them with shirking their duties, African American farm women exhibited patriotism through participation in wartime campaigns to produce food. When they canned thousands of jars of vegetables for the war effort, black women staked their claim to citizenship and showed that they deserved government services. "The colored people of the County have never been worked with before," reported Dazelle B. Foster from Davidson County. "They seem very anxious and appreciative of some one to work with them. The days and hours are not long enough for me to give the help called for."[21] Could not women who displayed such loyalty to the state ask that the state extend more of its resources to them in return?

When emergency federal support evaporated after the war, black home agents evaporated, too. Determined to keep the black agents employed, however, Jane McKimmon began to search for new funds. By 1922, she secured enough money to hire Emma McDougald and five other women as the first full-time black agents supported by the extension service. Yet by some measures rural black women had won the battle but lost the war. While the GEB had subsidized some forty Jeanes teachers for summer work with Homemakers' Clubs, the philanthropy diverted its resources to other projects once the extension service assumed responsibility for home demonstration among African Americans. Rural school supervisors continued to collaborate with white agents, but now they did so as "voluntary colored county workers." In 1920, for example, the white agent in Beaufort County shared publications, supplies, and the front seat of her car with the Jeanes teacher, who was "much interested in the uplift of her race" but who now gave demonstrations without compensation.[22] Until the mid-1930s, the number of black home demonstration agents never rose above seven, and the number of black farm agents fluctuated between fourteen and twenty. While black women and girls in a few counties enjoyed the services of a full-time agent, others had to settle for occasional instruction from black male agents or, after 1925, infrequent visits by an overworked black district agent. In the meantime, black agents joined a small group of African American professionals who sought to convince North Carolina officials to help meet the pressing needs of members of their race.[23]

IN THE FACE OF limited state support, black women used community networks, especially churches, to pursue their mission. "I have the gospel of good homes to preach to the people," Sarah J. Williams told ministers in Beaufort County in 1923, "and I want to know if I can do it from your pulpit." They never refused her. In 1924 Dazelle B. Foster visited a Wake County church where she hoped to organize a woman's club. It was Communion Sunday, and as the service wore on she nearly abandoned "the hope of being given a chance to speak." Finally, her turn came and she explained her plans, garnering the support of "the local leader," who instructed Foster to order a pressure canner so the churchwomen could preserve food. Congregations in Wayne County "mothered" home demonstration work, according to Emma McDougald, and church sponsorship secured "the support of the best people."[24]

A minister's backing helped agents win the confidence of rural African Americans who mistrusted representatives of a government that codified racial inequality and denied them rights of citizenship. As they went about their work, agents personified the state as a positive force in black communities. "The people seem perfectly delighted that the Government, State and Federal, should think enough of them to send one of their own women to work with them," Emma McDougald wrote in 1923. "Before, the Government was just a name to most of them, but now they feel that it is a real friend." Some black women regarded agents as, quite literally, a godsend. A Johnston County woman summed up her appreciation of agent Lucy Wade's lessons by declaring, "We give God and the State praise for sending her to us." Only divine intervention, she hinted, could account for such good fortune.[25]

The Johnston County woman gave thanks to state officials who often doubted that African Americans were worthy of the services they provided. In 1929, social workers investigating Negro child welfare in North Carolina included a survey of "race attitudes of county officials." The results are stark reminders of how pervasive and pernicious race prejudice was. Researchers included farm and home demonstration agents among the public officials polled as they gauged opinions on black education, land ownership, and voting. Assumptions of white superiority prevailed. Officials argued that most African Americans were too lazy or too stupid to take advantage of any opportunities that the government chose to bestow. While some officials favored blacks enjoying the same educational advantages as whites, most thought instruction should stop with the seventh grade or below because schooling made blacks "too biggety," "too uppity," and "harder to control." Most approved of blacks owning land, but few endorsed blacks voting.[26]

White home demonstration agents shared common stereotypes about African Americans. But if there is a silver lining in an otherwise dark cloud, their prejudices were slightly less vicious than those of other officials. For example, 30 percent of home agents favored equal educational advantages for black and white students—far more than the 9 percent of farm agents who expressed this opinion and more than all other officials except welfare superintendents. A third of home agents were willing for "educated and exceptional" blacks to vote; again, only welfare superintendents were so favorably inclined.[27]

When interviewers asked them to explain their opinions, the home agents revealed the depths of their prejudice but also offered glimmers of hope that a white supremacist ideology could be softened. The survey reports contained both direct quotations from informants and paraphrases of their comments. "The negro race is inferior to the white race and always will be," asserted the agent in eastern North Carolina's Washington County. "He should be given some opportunities to develop altho he can never develop to the extent that the white man does. He will always occupy a lower social and industrial plane." In the southern Piedmont, the Richmond County agent explained why she sponsored few demonstrations among black women. "They are too lazy and uninterested to take advantage of the opportunities offered them," she reasoned. Another agent asserted that African Americans "were created an inferior race and will always be cooks [and] house-maids. Give them a little education and you ruin them for the kind of work they are intended for. They all want to be school teachers."[28]

A few white agents expressed more moderate opinions. The Johnston County home agent used white self-interest to justify services for blacks. "'The negro is with us,'" she told interviewers, "'and anything that we can do for him will work to the advantage of the whites.' They are taking advantage of the opportunities offered them." Although the Hertford County agent thought that the "negro is racially inferior and will never equal the white man," she nonetheless favored "being fair to him and giving him a chance. The educated ones that she has come in contact with have not been biggety but have been humble and seem to know their place. They are not seeking social equality." In Martin County, the white home agent approved of blacks learning about "better sanitation, better housekeeping and better cooking"; she offered instruction by inviting white club women to bring their black cooks to a demonstration. "They are great mimics," the agent mocked, "and learn more quickly this way." Yet after the demonstration at least one black student had taken the initiative to turn teacher, sharing what she had learned about making sausage with neighbors. In fact, the Martin County agent reported that black women of the "better class" were asking for an agent of their own.[29]

Emma McDougald, the only black home agent questioned for the survey, countered the assumptions of her white colleagues. "'The negro,'" she assured the interviewer, "'is taking advantage of the opportunities offered him by the county.'" Improved homes represented just one of McDougald's accomplishments. Furthermore, she advocated academic courses that prepared black students for college because she worried that vocational education limited their options. "'The college woman or man,'" McDougald observed, "'is the one that gets along these days.'"[30]

The comments of white officials illuminate the narrow boundaries within which black home agents worked. Some white colleagues might already consider them "biggety" and "uppity" because they had attended college and held professional positions. Many white officials judged the very work the agents performed wasted on rural black women, unless it prepared them to be better servants for white families. As a consequence, black home agents performed a delicate balancing act. They had to prove to white officials that blacks "deserved" public services to which African Americans believed they were "entitled."[31] They acted as advocates on behalf of constituents who had little reason to assume the state's benevolence and served as vital links in an otherwise weak chain of public welfare for African Americans.

Within the context of white racism, the reports that black home agents submitted each year can be read as political tracts. Agents understood that poverty created severe handicaps for their constituents. When they described what might seem like painfully small accomplishments with pride, they also showed that black women were not lazy, were not content with their ramshackle houses, were not indifferent to their own well-being. For example, after encouraging "beautification of home grounds" projects among Beaufort County women, Sarah Williams discovered that "Lack of Funds and being tenants" had thwarted their hopes of making improvements. Nonetheless, many women searched the woods for shrubs and vines to replant in their yards and trained roses to cover "unsightly" porches, chimneys, and fences. "Almost every rural yard has some kind of flower garden," Williams noted, "and though it may not be what the landscape gardener would ask, the flowers lend beauty to the yard and house." In 1924 Dazelle Foster observed that Wake County women had accomplished little in the way of furnishing and decorating projects. "[O]wing to time and poverty," she explained, "my people have been kept from doing what they might have done." The next year, however, thirty women entered a kitchen improvement contest. The kitchen that won second place "seem[ed] to cling more closely" to Foster than all the others because its occupant had accomplished so much with so little. The woman lived in a three-room log house "way off the road, down a rough path . . . then up a muddy hill." Foster had doubted the kitchen had "the slightest chance of a prize." But the woman cleaned the walls with white clay, laid linoleum on the floor, installed screens and doors, and covered the table with oil cloth.[32] Such heroic efforts belied demeaning white stereotypes of African Americans.

As they worked with poor women, black home agents labored under material handicaps of their own. In 1925 Jane McKimmon appointed Dazelle

"Negro home demonstration members making box furniture," Duval County, Fla., 1930. (S-13568-C, Record Group 16-G, U.S. Department of Agriculture, Box 219, Home Demonstration Work—2 folder, National Archives and Records Administration, College Park, Md.)

Foster Lowe, who had recently married, the supervisor of black home agents. From her headquarters at the state's black land-grant school, North Carolina Agricultural & Technical College in Greensboro, Lowe administered a scattered district that stretched from the Piedmont to the coastal plain, one based on race rather than geography. As Lowe visited agents, she often found them working in bare offices without basic supplies or clerical help; she heard them describe going about their jobs bereft of canners, measuring spoons, or the most rudimentary equipment that county commissioners routinely provided for white agents. Pleas for better working conditions became a routine part of Lowe's annual reports.[33]

In addition, black home agents sometimes did double duty because few counties agreed to support both a farm and home agent to work with black families. Emma McDougald, for example, perennially complained that she worked alone in Wayne County. She took it upon herself to invite black farm agents from neighboring counties to conduct demonstrations and occasionally taught some classes on her own. McDougald insisted that "our men are anxious to know how to be better farmers," and in 1924 she grumbled that there was no male agent to help them. Two years later McDougald's appeals for a male agent grew more emphatic. "I realize that I cannot do much" for the men, she admitted, "only encourage them to hold on with the hope of a stronger arm to rescue them." Although McDougald had "studied agriculture in books," she had "never plowed a furrow." "One of our sorest needs," she reiterated, "is a Negro farm agent." While the need remained unmet, some men wanted instruction so badly that they attended women's club meetings and asked the home agent for help. McDougald did the best she could, "but that is not good enough," she observed. "We want a man, if you please, who has been trained to work with men and has time to give to the men." By 1929 McDougald worked Sundays and holidays to meet the needs of men, women, and youth. "The Agent very earnestly recommends that a Negro farm Agent be put in the field," McDougald petitioned once again. "The men's eyes are become open to their needs and so frequently call on the Agent to help them."[34] But no one answered her call for help.

Black home agents occupied a middle ground as they pressed for more resources, registered complaints about constraints imposed because of race and gender, and then made the best of what they had. Black agents preached a "politics of respectability" whose tenets appeared to pose little challenge to white authority. Mindful of the demeaning, racist assumptions that came as second nature to all too many whites, black agents encouraged their constituents to cultivate practices that contested and subverted those assumptions. Clean homes and yards, good grooming and attention to personal hygiene, and decorous behavior could translate into assertions of dignity and pride.[35] Such actions announced, "I am somebody."

As they went about their work, black home demonstration agents collaborated with clubwomen—be they poor tenants or landowners—to fashion programs that served political as well as personal ends. What appeared to be innocuous lessons in nutrition, sanitation, and household management became part of an assertive strategy to obtain better living conditions. Helping women grow more of their own food, for example, weaned them from dependence on commodities purchased on credit at inflated prices. Home-improvement campaigns provided occasions to press landlords for repairs to dilapidated houses. Encouraging women to keep household accounts prompted them to double-check a landlord's figures come settling-up time in the fall. Participating in a home demonstration club inspired interest in the world beyond the neighborhood and cultivated the skills that one needed for leadership.[36]

The efforts of black agents to promote good health among rural African Americans underscore how they connected poor constituents to public resources. African Americans were sicker and died younger than whites. In North Carolina, the infant mortality rate for blacks was double that for whites, and the maternal death rate among black women was higher as well. After studying "the neglected Negroes in rural communities," historian Carter Godwin Woodson concluded in 1930 that they "have remained embalmed in their ignorance of the laws of health." Inadequate schools, poor housing, deficient diets, and the absence of public health care created this "ignorance." As late as 1940, there were 312 white and 33 black public-health nurses in North Carolina; those numbers worked out to one white nurse for 7,974 whites and one black nurse for 32,302 blacks. There was one white doctor for every 1,127 white people, and one black doctor for every 6,499 black people.[37] With black health professionals few and far between, African Americans suffered more than their share of illness.

Black home agents joined rural communities to the national Negro health movement. This movement traced its origins to African American club-women's sanitation campaigns in the early twentieth century. In 1915, Booker T. Washington started National Negro Health Week to draw attention to the health-care needs of African Americans. Although men represented the movement nationally, women formed its backbone at the grassroots level. Advocates followed twin paths in their pursuit of better health among African Americans. First, they encouraged blacks to take personal responsibility for their own well-being; then, they prodded public officials to extend health services to African Americans.[38]

Through home demonstration clubs, black agents linked their work to this nationwide, indigenous mass movement. During National Negro Health Week in 1931, members of thirty-two communities in Columbus County cleared their yards of rubbish, destroyed breeding places for flies and mosquitoes, and built sanitary toilets. Under the guidance of agent Sarah Williams, ministers preached "health sermons" that enjoined worshipers to care for their bodies as well as their souls, and principals and teachers orchestrated school clean-up days.[39]

Black home agents acted as health educators, paying particular attention to pregnant women and children. Emma McDougald used churches to distribute bulletins on childcare to Wayne County's midwives and mothers. In Mecklenburg County, Wilhelmina Laws visited expectant mothers and instructed them on diet and neonatal care. "In nearly every instance," Laws "found that here was where much superstition prevailed." Agents solicited help for sick women too poor to help themselves and directed constituents to whatever public-health services counties offered. On her rounds in Wayne County, for example, McDougald discovered a couple whose ten-month-old daughter weighed only twelve pounds. After much coaxing, she convinced the parents to take the baby to the county health department for diagnosis. The undernourished child needed to drink more milk, but such a simple remedy lay beyond the means of the couple that had neither a cow nor enough money to buy milk, and poor diet reduced the mother's ability to breast-feed. McDougald intervened and convinced a local company to donate a regular supply of milk. "When the mother started to picking cotton and had to leave the baby with the other children," McDougald observed, "the child lost two pounds, but soon gained it back." By the time the baby celebrated her first birthday she was plump and cutting teeth. McDougald also helped Mattie Dickenson, a "very poor" mother of nine, save her six-month-old triplets—Ennis, Dennis, and Lennis—by obtaining food and free medical care for them. The agent did not stop there; through McDougald's intercession, several mothers with underweight and sick babies would visit "the Health Department where they find willing help."[40]

Agents introduced women to modern health practices and promoted new ideas about disease prevention. "The people are sorely neglected," Wilhelmina Laws observed in 1937. "A deplorable number still believe in old remedies that have no scientific bearing on promotion of health." Only recently had mothers allowed their children to be vaccinated or to have "other defects corrected." Of course, one reason that inoculations seemed so foreign was that public health care was virtually nonexistent. Nonetheless, when Craven County hired a black public-health nurse, agent Marietta Meares persuaded men, women, and children from the settlement of Mile Oak to receive the typhoid vaccine. The black nurse traveled to the community, set up a clinic in a grove of shade trees, and gave shots to 175 people as they filed by. More than 100 returned for a full course of inoculations. The Rowan County agent secured the services of a public-health doctor long enough for him to examine nearly 1,000 people, vaccinate more than 700, and diagnose "much malnutrition." Bertie County agent Lillian H. Andrews encouraged families to take advantage of clinics that the county health department sponsored. As a result, nurses vaccinated 445 children for smallpox, 228 for typhoid, and 56 for diphtheria.[41]

THE GREAT DEPRESSION exacerbated the marginal existence of black farm families and made the work of home demonstration agents even more critical. Federal money appropriated to meet emergency conditions increased their numbers and enhanced their work. In 1933 Jane McKimmon added eleven black and twenty-eight white emergency home agents to her staff. Just as they had been called into action during World War I, home demonstration agents of both races became adjuncts to county welfare officers as they distributed free seeds to relief clients and taught them to grow and can vegetables. Agents broadened their reach by training women—often home demonstration club members—to work with poor families who had to display a willingness to help themselves before they qualified for public assistance.[42]

Agents coordinated private and public relief efforts. In Columbus County, for example, Sarah J. Williams distributed flour donated by the Red Cross,

"Negro women's canning club with an exhibit of their canning, North Carolina, May 1932." (S-15909-C, Record Group 16-G, U.S. Department of Agriculture, Box 184, Food Preservation–Home folder, National Archives and Records Administration, College Park, Md.)

solicited canned goods and clothing to share with the poor, and convinced twelve midwives to deliver fifteen babies for free. By the spring of 1933, black Columbus County farm families desperately worried about how to feed themselves and pay their rents. Wages were so meager that they earned a mere penny a quart picking strawberries at local truck farms. Determined to give families a hand up, Williams took the message of "a good garden for every family" from tobacco barns to church sanctuaries, and she organized canning clubs. As a result, nearly 900 families raised gardens and 384 families on relief canned 16,490 containers of fruits, vegetables, jams, and jellies.[43]

In counties with no black agents, white home demonstration agents sometimes crossed the color line to work with black relief clients. The experience was often transformative for white agents, who witnessed poverty more grinding than they could have imagined, unfair practices of welfare officers, and the lengths to which black women went to help themselves. The white home agent in Johnston County observed the unrealistic expectations that the relief director placed on black clients and sympathized with their reluctance to cooperate. When the relief officer and home agent sponsored a meeting where the "colored women" were to bring their vegetables for canning and the home agent was to supply the equipment, the relief officer stipulated that the women "were supposed to give back half of their products to the Welfare chairman. Naturally," the home agent observed, "these poor colored women having no cars and no way to get around could not walk and carry their tomatoes and things up to the school house nor would they want to give away half of the meager amount they had, so nobody came." The white emergency home agent in Chatham County initially judged a black mother of three who cared for an invalid a poor prospect with whom to work. Yet the relief client attended canning demonstrations and eventually preserved twenty-two quarts of food, and in the process won the white agent's admiration.[44]

Some white agents considered their relief work among African American families the most satisfying service they had ever rendered. Eugenia Patterson joined the extension service as home agent in Washington County in the early 1930s. A native of the Piedmont, Patterson had never traveled east of Raleigh at the time of her appointment. When she visited tenant families on a plantation near Lake Phelps, their living conditions shocked her. In one home, a black woman suffering from pellagra occupied the sole bed that her family owned—a spring coil covered with burlap sacks—while her children slept on a pile of wheat straw. Tenants dipped herrings during the spring fish run and salted them for later use to supplement the fatback, meal, and molasses available from the farm commissary. Patterson convinced the farm manager to allot space for gardens so that tenants could grow more nourishing food. During August she conducted canning schools "for the colored people and such enthusiasm and response hasn't ever been met with by this

Agent." Patterson believed that she had opened up a new world "for these pauperized people, and a new one for Home Demonstration Work. We have touched the people who have needed it the most."[45]

As white and black agents experimented with relief work, Jane McKimmon decided to conduct a bold experiment as well. In 1933, she invited black home agents to join white colleagues at the annual staff meeting for the first time. The next year all extension agents—black and white, men and women —met together. The interracial meetings, McKimmon believed, "engendered a feeling of mutual respect" and created a setting where white agents could realize that "the Negroes by tact and real worth made a place for themselves."[46]

District agent Dazelle Foster Lowe recognized a window of opportunity, and she pressed for more concessions and recognition for black agents. In 1934, for example, she suggested to McKimmon that white subject-matter specialists should share lessons on nutrition, clothing, housing and other topics directly with black agents rather than working only with her and then expecting her to relay instructions to women under her supervision. In 1935, North Carolina extension officials appointed the first black subject-matter specialist in the South when they promoted Mecklenburg County agent Wilhelmina Laws to the job.[47]

Sensing a chance to increase her staff, Lowe pressed to secure permanent appointments in counties that enjoyed the services of emergency black home agents. It was an uphill battle, and she must have felt like Sisyphus. In 1934 Rowan County commissioners agreed to retain the black agent after Lowe visited them four times and enlisted the white agent's support. The commissioners had set aside money for the white agent to hire a stenographer. The sum was insufficient, however, so the white agent suggested that the money be used to complete the black home agent's salary. Such local gestures were exceptional, however, and in her 1934 annual report Lowe recommended "that some plan be worked out whereby the State can be given the power to appoint Negro workers in counties where most needed as sentiment is not always favorable." Black families comprised a quarter of the population in fifty of North Carolina's one hundred counties and therefore warranted home demonstration agents, according to the extension service's formula, but only eight employed them. The need was especially acute in east-ern counties with high proportions of black tenants. Lowe suggested that county commissioners resented a black representative of the state meddling in local affairs when she concluded, "Back of it all seems to be prejudice and fear of the 'big landlord' enlightening the tenant." Nonetheless, by 1936 North Carolina extension officials used state and federal money to hire ad-ditional black home agents, raising their number to twelve, and a few more counties added agents each year thereafter.[48]

Ironically, as the number of black home agents gradually rose, the number of black farmers began to fall. The same agricultural bureaucracy that mod-estly increased its support for extension work among African Americans during the 1930s designed policies that worked to the detriment of thou-sands of black tenant farmers. The Agricultural Adjustment Act (AAA) of 1933 paid farmers to reduce the acres of tobacco and cotton grown. Fewer acres required fewer tenants, and, rather than sharing government subsidies, landlords often pocketed the proceeds and let displaced workers shift for themselves. As a consequence, thousands of landless black farmers joined the dispossessed, looking to the "big landlord" in Washington for help. Between 1930 and 1940, the number of black farmers in North Carolina dropped from 74,636 to 57,428. Of the survivors, 75 percent were tenants.[49]

Across the South, displaced farmers protested the consequences of New Deal policies. Some mounted individual fights by writing letters to adminis-trators in Washington who they hoped would intervene with their landlords and help them retrieve their fair share of AAA payments. Others joined col-lective protests. The loudest cries came from the cotton fields of eastern Arkansas where sharecroppers complained of wholesale evictions and fought back through the Southern Tenant Farmers' Union. Although AAA adminis-trators denied the legitimacy of the grievances, they could not ignore the protests altogether.[50]

By 1936 the USDA and the Agricultural Adjustment Administration had to respond to challenges from all sides. That year the Supreme Court ruled the first AAA unconstitutional. Its replacement, the Soil Conservation and Do-mestic Allotment Act, contained provisions that sought to channel payments to tenants and to halt evictions. In addition, black farm-extension agents began to press for more information about the new AAA policies and for more black USDA employees to get that information to black farmers. At a March 1936 soil conservation meeting for USDA and AAA personnel in Memphis, veteran extension agents Thomas M. Campbell of Tuskegee and John B. Pierce of Hampton, along with other black professionals, petitioned policymakers. They suggested that the USDA organize educational conferences on the new legislation, increase the number of black extension agents, advise teachers at black land-grant colleges about the new allotment and conservation act, and appoint African American representatives to state and local boards that settled complaints. Finally, the petitioners asked "that the rural Negro women be given a definite place in the new program."[51]

Cully A. Cobb, administrator of the AAA's Cotton Section in the lower South, caught the brunt of the criticism. A Mississippi native and loyal ally of planters, Cobb nonetheless had to make some concessions to African American dissenters. In the fall of 1936, he named Claude A. Barnett, the Chicago-based president of the Associated Negro Press, as a special assistant, no doubt to quiet the Republican editor's criticisms of the AAA and to win the favor of African Americans in the North as the elections approached. At the same time, Cobb appointed three black special AAA field agents to travel the lower South, apprising black extension agents of program guidelines as they changed from year to year, and building support for the program among black farmers. Cobb chose Albon L. Holsey, director of publicity at Tuskegee Institute, James P. Davis, president of the National Federation of Colored Farmers, and Jennie Booth Moton, former first lady of Tuskegee and veteran clubwoman who had retired with her husband, Robert Russa Moton, to their home in Tidewater Virginia.[52]

Jennie Moton's particular task was to "contact Negro farm women in the southern region in order to get their viewpoint on just how the Agricultural Conservation program is working out." She accomplished that assignment by traveling with home demonstration agents and visiting churches on Sundays. As she investigated, she also proclaimed "the good news and glad tidings of the Agricultural Conservation Program to innumerable groups."[53]

Jennie Moton was both an odd choice and a perfect choice for the job. Although she had directed Women's Industries at Tuskegee after the death in

1925 of Booker T. Washington's widow, Margaret Murray Washington, she had little experience working directly with rural women. In fact, Thomas Campbell had suggested that Cobb appoint the home demonstration agent from Macon County, Alabama, to the position. Yet Moton brought to the job contacts with influential African Americans all over the country and with the white liberal establishment of the South, and there was little chance that Moton would use the position to demand far-reaching reforms. Although active in black women's clubs at regional and national levels and a member of the Commission on Interracial Cooperation since its founding in 1920, Moton had never emerged as a strong leader or effective spokesperson. Perhaps operating in the shadow of Margaret Washington had stunted her growth; maybe she was a reluctant reformer, cast in a role for which she had little talent or interest because of her marriage to a prominent educator.[54] No doubt, politics also influenced her selection, as Cobb sought to squelch criticism of the AAA's deleterious effects on black farmers. Moton's appointment would gain more attention than that of an obscure home demonstration agent who might be more qualified for the task. Indeed, Holsey, an old friend and former colleague of the Motons, had helped to engineer her selection. When he informed Barnett of her new post, the editor commented, "I am sure that she will get a lot of satisfaction out of being in active life again."[55]

During the next five years, Jennie Moton dedicated herself to the AAA work and finally came into her own. Her five children were either married or nearly grown; and her husband's retirement meant that she no longer had to serve as Tuskegee's first lady. In 1937, members of the National Association of Colored Women (NACW) elected her to the first of two terms as president. Wearing two hats, she maintained an exhausting schedule as she traveled constantly on behalf of the AAA and the NACW.[56]

It is easy to count the miles that Moton logged as she crisscrossed the region from South Carolina to Georgia, from Mississippi to Arkansas, but it is difficult to discover what she said and what she did once the train pulled into the station. She left behind no texts of speeches delivered, no lengthy reports on her findings, no recommendations about how AAA policies might be amended to better meet the needs of rural black women. Most of her letters consist of thanking hosts for their hospitality during recent visits and informing them of her hectic schedule. Rarely did Moton display any particular sympathy with rural blacks; on the contrary, she could be downright disparaging as she invoked images of black women who cooked better "after they have broken a door off the stove" and of black men who tinkered with their worn-out cars right on the side of the road.[57]

Jennie Moton nonetheless carried weight as a figurehead. After conferring with Moton, fellow field officer J. P. Davis praised the AAA's "efforts to improve the home life of the colored farm mother and her family. . . . The first time in history of the career of our racial group has this consideration been accorded us," and he was grateful that she had "decided to tackle this grave problem." The black state agent for Georgia told Moton that agents were heartened by the fact "that a colored woman had been given such a high position with the A.A.A."[58]

Moton's visits also boosted the morale of weary black agents. "Our workers," explained a black farm agent, "coming in here from the field, coming from the scenes of many difficulties and problems, needed much the kind of inspiration which you brought." A district home agent in Oklahoma confided, "Sometimes it seems that our efforts are so in vain, and when you come or write it gives inspiration and the feeling that after all, what we do is worthwhile." Other agents, without access to the information loop controlled by white agriculture officials, appreciated the opportunity to learn more about AAA regulations. "Unfortunately, we have not participated in this program at all," an Arkansas district agent informed Moton, "nor have any of our staff of Negro Home Demonstration Agents been called in for any meetings or conferences pertaining to the Agriculture Programs and for this reason . . . our knowledge of the program is very limited."[59]

When Moton visited extension agents, she shared information about AAA policies and encouraged farm women to plant gardens as their contribution to soil conversation. To see policy in action, she often accompanied black home agents on project inspections, where she praised pantries lined with home-canned fruits and vegetables and bedrooms remodeled on a shoe-string budget. Complimenting agents' work within earshot of local white officials was important because each year county commissioners had to allocate a portion of the black agents' salaries. Moton also relayed her admiration for the home demonstration agents' work to superiors in Washington. "I am constantly telling someone of the splendid work [black home agents] are doing," she told the black district agent in Oklahoma, "and I am as frequently asking that the number of workers be increased."[60]

Black advisers to federal farm officials placed the task of increasing the number of African American extension agents at the top of their policy wish list. During 1938 and 1939, for example, Claude A. Barnett toured the South on behalf of the AAA to evaluate how black farmers fared and to promote AAA policies. He contended "the name 'Washington' has taken on a new significance to tenants and sharecroppers. They realize even though they do not know how to take advantage of it there is a power in 'Washington,' which even their bosses respect." To help farmers even more, Barnett advised the extension service to hire additional black farm and home demonstration agents and suggested that the state AAA administrators hire black aides. The more black farmers learned about their rights under AAA, Barnett reasoned, the more they could take advantage of them. In short, the best way to help the poorest black farmers was to increase the number of black professionals to work with them. After having been "extremely conservative in its approach to the Negro farmers through the years," as Barnett put it, the U.S. extension service was being held accountable.[61]

North Carolina's extension officials were being called on the carpet, too. In what appears to have been an orchestrated campaign, within a matter of weeks late in 1939 several North Carolinians wrote state and federal extension directors to request information about the number of black agents or how much money the USDA invested in black land-grant colleges or the number of black vocational agriculture teachers in the public schools. Elizabeth L. Brown of Edenton, C. H. Ford of Roxboro, Thomas E. Moultrie of Dunn, Mott Redfearn of Marshville, Raymond Perry of Wendell and his neighbor Carl Atvie Perry—all wrote to ask what the USDA was doing for African American farmers.[62] Responding to pressures from Washington and pressures from the grass roots, North Carolina extension officials did hire more black agents. Between 1935 and 1943, the number of black home agents in

North Carolina climbed from twelve to twenty-five, and the number of farm agents grew to thirty-eight.[63]

Black home agents epitomize the paradox of depression-era agricultural reforms. Policies that hurt the poorest farmers generated grievances that policymakers addressed by hiring more black extension agents to work with the survivors. Black home agents like Dazelle Foster Lowe shrewdly played one level of government off against another to wrest better treatment for local agents and more services for African American communities. By 1938 delegations of "key men and women" from four counties in eastern North Carolina with large black populations had appealed to Lowe and their commissioners for home demonstration agents of their own. In the meantime, Lowe had convinced Ruth B. Current, the state agent who succeeded Jane McKimmon, to provide paid summer leaves for black agents eager to continue their education. Lowe also had determined how to influence county commissioners as they deliberated personnel decisions. She first nominated a woman she felt best qualified to serve as home agent, and then she sought commissioners' advice because she realized "it tickles the fancy of the County Boards if they are consulted."[64] This strategy allowed Lowe to hire the woman she wanted even as it maintained the illusion of white male control.

Lucy Hicks Toole was among the new home demonstration agents hired in 1936. A 1933 graduate of Winston-Salem State College who had taught high school home economics for a year, she accepted an assignment in Johnston County. Because most of the farm women she served depended upon whites for employment, Toole realized that she would need to combine advocacy and education. Her persistence paid off. When some of her clubwomen complained that leaky roofs, unfinished walls, and rickety steps impeded kitchen improvement plans, Toole "took down each lady's name and things which needed" fixing, visited their landlords, and extracted promises that they would make repairs. A month later Toole followed up her initial visits, reminding landlords of their pledges. Much to the surprise of the club president, the man who owned her house provided enough lumber to ceil the kitchen and bedroom, "told her husband to get some one to help him put it in," and furnished paint for the two rooms. Another landlord required more prodding before he mended sagging floors and drafty walls. A month after Toole's initial visit to him, "she went to see what he had done and that was nothing." The landlord attributed his tardiness to a failing memory; after Toole's reminder, he supplied the materials needed for the repairs.[65]

The agent's memory was not nearly so short as the landlord's. Sixty years later, at the age of eighty-eight, Lucy Toole DeLaine recalled the recalcitrance of white landlords as if she were still negotiating on behalf of tenant families. The story of her struggles with one prosperous—and obstreperous—landlord and merchant typifies the persistent diplomacy necessary to improve housing conditions. The first time she visited the landlord at his office, DeLaine recalled, she described houses that left tenants shivering during the winter "even with a roaring fire." Would he provide materials so the tenants could patch the cracks in walls and floors? she asked. "Make it warmer?" the owner answered incredulously. "They're not making me enough money for me to do that." DeLaine appealed to his self-interest. "Maybe if you help them inside the house," she suggested, "they would do a better job on the outside." He insisted that he could not afford to improve the houses. DeLaine proposed that he let tenants use cardboard boxes discarded at his store to insulate their walls. How would the tenants get the boxes from his store to their houses? he mused. DeLaine suggested that he could take the boxes on his weekly tours to inspect the farm. "I don't have time," he snapped. "I am a busy man." DeLaine stood her ground. "Sometime if you help people, show them that you are interested in them, they do better," she observed. "I'll see," he replied. Eventually, the tenants received cardboard to line the walls of their houses.

One victory to her credit, DeLaine began cajoling the landlord to install window screens in the tenants' houses. "They don't need no screens," he said. But "the flies in the summer are terrible," DeLaine told him. "They light on the food." The tenants could fan the flies away, the landlord replied. "I am quite sure you don't have any flies in your house light on your food," DeLaine rejoined. No, he did not, he agreed. Again, DeLaine appealed to his self-interest. "The health conditions of the people have a lot to do with how they work," she reminded him. "So I stayed on him about screens." Months later, the landlord put them in. Finally, DeLaine waged a third protracted but successful campaign to convince the landlord to build sanitary outhouses for tenants who otherwise relieved themselves in the woods.

From the vantage point of vigorous old age, Lucy DeLaine acknowledged the courage required for a young black woman to take on a powerful white landlord with her insistent appeals. "It's a wonder they hadn't strung me up," she reflected. "It was like pulling eye teeth. You just worked on them one thing at a time. You couldn't push too much. So that was from one landlord to the other. You just plead, plead, to do something for this person. He's giving you his best. Meet him half the way. But that was a hard thing for [the landlords] to see. They felt that they needed it all, you didn't need anything. I guess I was as tired when I left extension as though I had hammered all day. But I got some things done."[66]

Ever mindful of their constituents' poverty, black home agents improvised and invented ways for women to make a little go further. A recipe that black agents concocted captures their ingenuity. A cheap cut of pork known as fatback was a staple of poor black southerners' diet. High in calories but low in nutritional value, fatback was such a symbol of poverty by the late 1930s that some children cringed with shame when they packed it in their school lunches. Agents created a dish they called country style salt pork. To prepare it, the cook dipped the fatback in eggs, dredged it in cornmeal, and then fried it. Frying would appear to compound the harmful effects of the fat meat, but a batter of eggs and meal enhanced its nutritional value. The new recipe, moreover, camouflaged fatback's appearance and taste. Clubwomen who served country style salt pork enjoyed a good laugh when they tricked husbands into thinking they were eating fish or oysters. The dish, reported Bertie County agent Lillian H. Andrews, "created more publicity than any other lesson given. The home agent received quite a few bouquets even from the pulpit." As one minister who encouraged church members to take advantage of the extension service extolled, "'Why Miss Andrews can take fat back and fix it to taste just like fresh fish.'"[67]

It would be tempting to use country style salt pork as a metaphor for the work of black home agents, as a way to criticize them for concealing the worst symptoms of poverty rather than helping women effect real social change. Certainly, home agents counseled individual responsibility and placed tremendous stock in the appearance of homes and the women themselves. Most of the projects they sponsored were palliatives rather than cures.

Yet black home agents created social spaces where rural black women could develop a sense of accomplishment and self-worth. A story from Wake County in 1937 illustrated that a "politics of respectability" was more than window dressing. The agent began the anecdote by noting that one "can no longer pick out country women on account of their ill fitting clothes and unbecoming colors." Appearance could translate into power, the agent implied, as her story segued into an example of growing self-confidence among home demonstration club members. "The women have developed poise, self-respect and independence," the agent reported. "They are fast becoming leaders of affairs." In fact, Wake club members had appointed a three-woman committee and charged members "to attend any meeting held in the county of importance." After monitoring a state welfare meeting in Raleigh where the "Old Age" pension topped the agenda, the committee chair could answer clubwomen's questions about Social Security.[68]

Home agents continued to link poor black women to a growing number of government services. When Alamance County clubwomen gathered for their annual achievement days in 1940, the entertainment included soliloquies about mattresses stuffed with surplus cotton that the federal government made available. While white state agent Ruth Current described the speeches as humorous, black clubwomen might have interpreted their message to mean that the state that denied them so many rights could also befriend them. The first speaker personified "'the old straw tick. For years I have been used as a bed. I have caused people to roll and tumble all through the night. . . . I have had my day, but my time is not much longer for I hear they are coming with cotton to take my place.'" The second speaker symbolized "'the new government mattress. I only cost $1.00 and a little work. You know I get tickled at the miration some folks make over me. I am bringing a lot of good rest to poor families.'" After praising the mattress's comfort, the clubwoman reminded the gathering, "'I am available to all low income families.'"[69]

Black home agents built loyal followings among members of their communities. When government support for Emma McDougald's salary was threatened in 1932, the "Negro citizens" of Wayne County called "a monster mass meeting" to rally support and raise money to retain her. After organizing a County Agricultural Association, the group vowed to raise the $600 necessary to pay her. Led by a truck farmer, a physician, the superintendent of black schools, teachers, and pastors, the group passed a resolution asserting that home demonstration work was "the best agency in our county" and that "to lose that agency would be a tremendous setback to our agricultural progress." They pledged "themselves unqualifiedly now and for the future to see to it that the work, whether or not paid by County or State, shall not perish from the earth." The association made good on its promise. Within a matter of days, the group had raised $155 through one-dollar donations and delivered the down payment on McDougald's salary to state headquarters in Raleigh.[70]

Black home agents became trusted friends to clubwomen, and they assumed a multiplicity of roles. In 1936, Dazelle Foster Lowe declared that the black home demonstration agent was "a real legal adviser, social and spiritual worker in the eyes of the rural folk." Wilhelmina Laws amplified the description when she noted, "the home demonstration agent is called on for everything." Where there was no public-health nurse, the agent "conducted clinics, prescribed diets, and looked after the health of the people in general." She served as a welfare officer and employment counselor for clubwomen who needed jobs. "In counties where there are no high schools," Laws continued, "the Agent finds homes for the girls where they can go to the city high schools."[71]

By the 1930s, rural African Americans had learned that a growing welfare state was Janus-faced. It could give and it could take away; it could be friend or foe. Policies created at the federal level might be implemented locally in ways quite different from those originally intended. As women in the middle, African American home agents made the case that black rural women and their families deserved the same government services as whites and were entitled to a decent standard of living. By helping black women "make the best of what they had," they also gave them the confidence to believe that if one door was closed another might open—if they knocked loudly enough.

CHAPTER SIX

From Feed Bags to Fashion

IN THE 1970S, north Georgia students documenting traditional mountain culture for *Foxfire* magazine asked several rural women to describe sewing with feed sacks. Mary Franklin, Harriet Echols, and Lettie Chastain reminisced about turning the bags into everything from curtains, sheets, pillowcases, towels, and quilts to underwear, aprons, shirts, and dresses. "We just made do in the country," explained Harriet Echols, "you had to. Most everyone made clothes out of the bags. There was never a bag thrown away." For Echols and her neighbors, sewing with sacks was a quintessential symbol of farm women's thrift and resourcefulness. The Foxfire students and the women themselves placed the stories about feed bags within the context of a rural culture whose members had long believed that to waste not was to want not and framed them with memories of the depression.[1]

Yet the stories revealed as much about economic changes as about cultural continuities. The women noticed, for example, that the number of bags available for sewing multiplied as the poultry industry grew in north Georgia and that feed dealers began to entice customers by packaging their products in colorful print sacks. "I'd try to go buy feed so I could match the bags with what I had at home," Harriet Echols recalled. "I'd take a bag of every color that I had." Using feed bags to "make do" was as much a measure of how enmeshed the women were in a world of commerce as it was of how independent they were of store-bought fabrics. Although cast in memory as a symbol of a rural culture that valued meeting as many of one's needs as possible, sewing with feed bags was also a sign of rural women's incorporation into a national economy as producers and consumers.[2]

COTTON BAGS BECAME a source of fabric in the later nineteenth century, when they replaced wooden barrels as containers for staple goods such as flour. The Bemis Brothers Bag Company, founded in 1858 in St. Louis, was the first to produce textile bags. The company established its own cotton mills and bleachery, and by 1910 it ran eleven factories to supply the growing textile bag market. By the turn of the twentieth century, Fulton Bag and Cotton Mills of Atlanta dominated the bag market in the South. Some evidence suggests that women recycled flour bags into clothing in the late nineteenth century, and remnants can also be found in quilts dating from this period.[3]

During the 1920s and 1930s clothing made from cotton bags became an emblem of poverty, a testament to ingenuity, and a badge of pride. Depression-era photographs show the poorest of America's poor clad in sack garments. When writer James Agee and photographer Walker Evans traveled to Alabama to record the words and images of three white sharecropper families, they found the women wearing shapeless bag dresses. Farm Security Administration photographer Marion Post Wolcott pictured a black migrant vegetable picker in Florida using a cotton sack as a skirt and another worker with a large sack draped over her shoulders and back like a cloak.[4] On a lighter note, an amateur poet's ode to sack garments praised her mother's resourcefulness and bragged about wearing panties that still bore brand names and advertising that betrayed the fabric's humble origins. According to the ditty, "One pair of Panties beat them all, / For it had a scene I still recall / Chickens were eating wheat / Right across my little seat."[5]

Relief workers and home demonstration agents in southern states used cotton bags in their efforts to help poor rural women. Among the most successful projects of the Georgia Emergency Relief Administration were "classes for young girls who longed for pretty things but could not afford to buy them and could not make them." Relief workers taught the girls "to admire real beauty and cleanliness and to make the most of simple and inexpensive materials"; and dresses sewn from Dixie Crystal sugar sacks proved especially popular. When an Alabama home demonstration agent sponsored a sewing contest as part of a campaign promoting self-sufficiency, she promised a prize to the woman who made the best dress from a fertilizer, feed, or flour sack. The winner fashioned her dress from a Jazz-brand feed bag dyed a rose color, and her prize was a twenty-five pound bag of flour. In North Carolina, home agents sponsored dress revues where they confirmed that hard-pressed women did not have to sacrifice style.[6]

Transforming bags into a suitable fabric required painstaking labor. In the early 1930s, manufacturers packaged their products in white cotton sacks and used durable inks to emblazon them with colorful brand names and company logos. Women had to scrub out these designs before the material could be used. The inks, one North Carolina woman recalled, "could be removed if you had lots of patience and elbow grease." First, she dissolved Octagon soap and Red Devil lye in warm water and soaked the bags overnight. The next day she rinsed the bags, rubbed them on a washboard, and then boiled them in a wash pot to remove the obstinate dyes. After a final soaking in Clorox and several more rinsings to remove the strong bleach, she dried and ironed the bags and turned them into fabric that resembled muslin. The process of ink removal even entered southern folklore. Poor whites in Alabama believed that a full moon made the sacks release the inks more easily.[7]

Once they had bleached the bags, women altered them with natural dyes or embroidery. A north Georgia woman recalled that she boiled black walnut hulls and oak bark and then dipped white sacks in the water to absorb the brown color. Some of her neighbors boiled sumac berries and dyed sacks a reddish tint. Women with a talent for fancy stitching embellished the bags with flowers and other designs after they fashioned them into curtains, dresser scarves, or aprons.[8]

Although using textile bags for clothing is usually associated with scarcity and "hard times," women had to have enough money to buy the products that the bags held. "When the depression hit," recalled Lettie Chastain of north Georgia, "we were real poor." As a result, the family substituted home-grown cane syrup for store-bought sugar, thereby eliminating a source of sacks.[9] Women who had a large supply of feed bags belonged to families who raised poultry or livestock for market. In other words, for a woman to have access to a lot of "free" bags required a certain degree of affluence. A farm's livestock mix determined the kind as well as the quantity of bags available. For example, Edythe Hollowell Jones explained that she used bags in new ways after she married in 1941 and moved to an eastern North Carolina farm where her husband raised far more chickens than her mother had kept. Irene Barber Hollowell had used flour, sugar, and salt bags to make dishcloths and towels, but she had never raised enough chickens to justify buying hundred-pound bags of feed. Jones's husband, however, raised two hundred laying hens, and her mother-in-law "had been washing the bags that the [chicken] starter and laying mash came in and using the material for some time" to make sheets, mattress covers, and nightgowns as well as smaller household items such as dishcloths.[10]

Women who had surplus sacks turned them into a commodity. A Bladen County, North Carolina, woman who raised more than a thousand broilers in 1937 explained to a home demonstration agent the different ways in which leftover feed sacks entered networks of trade. She sold some to neighbors for cash, she swapped others for "help with my work," and the rest she used to make towels, aprons, and "other useful things for the home."[11] In the 1930s when Gerti Roberts's mother began raising broilers commercially in Wilkes County, North Carolina, she also joined a lively trade in feed bags. "Then a lot of folks [would] hear tell of your raising chickens," Roberts recalled, "and they'd come and buy sacks from you. They'd give twenty-five cents [apiece]" for bags that contained 1⅓ yards of fabric. At the time, Roberts worked at a hosiery mill in North Wilkesboro, where an elderly woman who swept the floors became a regular customer for her mother's sacks. "Mother would wash them out," she recalled, "and I'd take them up there. I don't know how many sacks I did carry up there and sell."[12]

The creative ways in which rural women used the bags and their desire for cheap cloth gained the attention of bag manufacturers, millers, and retailers. To capitalize on a grassroots market in sacks and appeal to women's resourcefulness, manufacturers turned out bags in an array of prints and colors. Although dating this development precisely is difficult, print sacks had clearly joined plain cotton ones by 1940. During the next two decades, buyers could pick and choose among bags that came in bold geometric patterns, simple stripes, modest checks, sedate plaids, and dainty florals.

The development of dress print bags also coincided with a shortage of certain fabrics for civilian use during World War II. Until the war, manufacturers packaged agricultural products in bags made of burlap as well as cotton. In 1941, however, the United States Office of War Production Management began allocating burlap, and by the next year all burlap went for military use. Needing another material for their containers, feed millers turned to cotton.[13]

Manufacturers of dress print bags had to accommodate women's desire for unblemished fabric and feed millers' desire to build brand loyalty. To meet these goals, bag manufacturers designed paper band-labels attached to bags with water-soluble glue. This innovation protected the cloth from inky stains and provided room for colorful advertising designs.[14]

The introduction of dress print bags indicated that feed dealers recognized women as farm decision makers and energized women's trade in bags. Although Arthur Fleming considered himself the manager of the broiler operation he started in north Georgia in 1939, he relied upon his wife and children to feed and water the flock. There were a lot of empty sacks to dispose of. Fleming's wife used some of the sacks to make dresses, but "what she didn't want she would sell. . . . There were people who came around and bought those sacks, and the wives would sell 'em. Then you'd have neighbors that come in, that didn't raise chickens, and they'd look at the sacks and they'd pick out what they wanted." Sometimes his wife had on hand as many as two hundred print sacks for sale.[15] Lettie Chastain's family did not raise poultry commercially, so she bought sacks from a neighbor who did. "The wife would pick out the one kind of whatever [print] I wanted," she explained. "She'd call me when she got a batch in, and I'd tell her which I

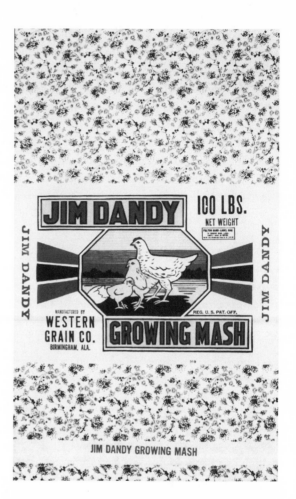

Dress print bag for Jim Dandy growing mash. (Photo courtesy of Office of Imaging, Printing, and Photographic Services, Smithsonian Institution, Washington, D.C.)

wanted."[16] After Ruby Byers's father started raising broilers in the late 1930s, she and her mother sewed with feed sacks and bartered the surplus with neighbors. "I know people would get so excited when a new bunch of prints would come in," Byers recalled. "And if you got a few that you liked but didn't have enough to make what you wanted, you traded bags" with a neighbor who had the desired print.[17]

Feed dealers testified that appealing to women bolstered their business. In a 1945 advertisement, the National Cotton Council urged feed manufacturers to use cotton bags, because "When women face a choice of brands, a cotton bag often swings the decision—for, like the product it contains, it has real utility in the home."[18] A feed dealer in 1948 complained, "Years ago they used to ask for all sorts of feeds, special brands, you know. Now they come over and ask me if I have an egg mash in a flowered percale. It ain't natural."[19] Other merchants reported that when farmers went to town to buy feed they brought swatches of cloth to match the prints that their wives and daughters wanted.[20] Edythe Hollowell Jones recalled that during the 1940s when a dealer delivered feed for her husband's chickens each week, she and her mother-in-law supervised the truck driver as he unloaded the bags. They surveyed the options and chose bags that suited their tastes and needs as they planned their home sewing.[21]

For women who controlled their own poultry operations, the feed bags served as a valuable bonus. For women whose men folks managed the enterprise, choosing the bags was one way to assert their influence in a changing farm economy, and selling bags replaced some of the income they had lost when men appropriated the chicken business.

TO FURTHER ENCOURAGE bag consumption, two trade associations—the National Cotton Council and the Textile Bag Manufacturers Association—initiated national advertising campaigns in 1940. For the next two decades, the associations aimed promotions at feed dealers, bag manufacturers, and farm women. They placed advertisements in trade journals, published booklets that instructed women how to sew with cotton bags, and sponsored sewing contests. They touted feed bags as a source of fashionable clothing and revealed the power women wielded as consumers.

The textile bag industry placed much of its advertising in the trade journals written for feed manufacturers. Bag companies promised that a shrewd marketing strategy would help dealers boost sales of feed grains. "Smart packaging is vital in the successful marketing of your product," advised the Percy Kent Bag Company. "Make sure it is the best dressed item on the retailers' shelves." "As a premium," the same company pointed out, "the Ken-

Print Bag is no added expense."[22] The Bemis Bag Company supplied retailers with posters for display in their stores.[23] The Chase Bag Company described its many colorful "patterns which assure repeat business and lasting demand for your products."[24] According to a 1948 advertisement aimed at feed dealers, "one of the most popular appeals today to thrifty rural housewives is the reuse of valuable cotton feed bags. This is of increasing importance to you because the more dress print bags that women demand, the greater the sale of your product—if it is packed in these colorful containers."[25]

One popular advertising strategy lay in a familiar appeal to rural women's thriftiness and self-reliance. A 1944 cotton council advertisement, for example, informed feed dealers that its latest campaign in farm magazines, teacher publications, and small-town weeklies showed women "how to make the material from cotton bags do double wartime duty." Rural women, the advertisement continued, "are emphasizing conservation through the reuse of cotton bags. Products packed in cotton bags pack a popular sales appeal that's all their own."[26] In 1945, the council claimed that more than one million women were using "A Bag of Tricks for Home Sewing," because "not only does it bring IDEAS, but it promotes THRIFT! . . . To women a cotton bag is not just a container. It represents a plus value for the product inside it."[27] The Fulton Bag Manufacturing Company promoted its bags in 1948 as "Budget-Savers!"[28] The same year the Chase Bag Company wrote that from its bags "Thrifty 'home-managers' the country over are fashioning many practical, attractive items for the home and family."[29]

By the late 1940s, however, advertisements began to emphasize the print bags as a source of fashionable clothing. In 1947 the Percy Kent Bag Company boasted A. Charles Barton as its design director and noted his connections with capitals of urbanity. "From his studio in New York, Mr. Barton, one of America's foremost designers, sends out the distinctive ideas for which P/K Bags are famous," claimed the advertising copy in *Feedstuffs*. "European by birth and education, he has won wide recognition . . . as one of America's foremost fabric designers." If Barton was a cosmopolitan designer, the publicity implied, rural women gained a sense of worldliness when they wore his fabrics. At the same time, rural women influenced textile prints. Barton had recently toured the American Midwest "to view first hand the many uses to which Ken-Print material is being put by the versatile homemakers of this area, and to get new ideas for future P/K patterns. Upholding the Percy Kent tradition of 'always something new,' he promises more of the clever, colorful designs that have made Ken-Print Bags the 'glamour sacks' of America."[30]

Farm women collaborated with fabric designers to define fashion for the Bemis Bag Company. In 1947, Bemis boasted, "Women go for these New York Fashions. Straight and fast into Bemis plants go these exclusive prints from the world's fashion center. And out again they come as Bemilin Dress Print Bags to carry your feed in a package women are eager to get. These patterns are usually reserved for luxury fashion only." A "noted New York designer creates them," the advertisement continued.[31] The next year, however, Bemis's promotional strategy did an about-face. "Who knows best what women want in fabrics for dresses, curtains, etc.?" its ads asked. "Why, the women who use these fabrics, of course." Bemis had "groups of typical farm women judge a wide range of patterns and pick those they like best." The company then adopted designs that this "Supreme Court of Farm Fashion" had chosen.[32]

Sex appeal furnished another promotional theme. A 1947 Percy Kent advertisement captured a woman dressed in a two-piece sun suit in a provocative pose, with one hand on her hip and the other behind her head. "Nice Package!" the copy line proclaimed. "Flour bag to sun suit is the short story behind this eye-appealing package."[33] The Erwin Manufacturing Company in 1948 ran a series of cartoons, all using ideas of beauty, sex, or romance to sell its product. One featured two buxom beauty contestants grumbling that a homely, flat-chested, bespectacled woman had won the crown. The blonde groused to the brunette: "With a woman judging the contest I knew that ERWIN print bathing suit would win!"[34] Another cartoon advertisement focused on a courting couple that had parked their car in front of the justice of the peace's house on a moonlit night. The woman, wearing a strapless dress, asked, "Are you sure you're not just temporarily dazzled by my ERWIN print dress?"[35] Marriage prospects were the subject of an advertisement in which a suitor climbed a ladder to a second-story window where his bride-to-be waited to elope. A feed store stood across the street. Leaning out of the win-

Percy Kent Bag Company advertisement: "Nice Package!," *Feedstuffs*, 1 February 1947, p. 67. (Photo courtesy of Office of Imaging, Printing, and Photographic Services, Smithsonian Institution, Washington, D.C.)

dow, the woman asked, "Can you pick up a few more ERWIN print bags at the feed store, Honey? I haven't finished my trousseau yet."[36]

Besides reminding feed manufacturers of their female customers, the cotton and bag industries appealed directly to women by distributing booklets that suggested a multitude of uses for textile bags—everything from dresses and aprons to shoe cases and yardstick holders. In 1940, the Textile Bag Manufacturers Association's idea-and-pattern booklet advised, "For Style and Thrift, Sew with Cotton Bags." By the end of the decade, the trade associations published "Bag Magic," "Thrifty Thrills with Cotton Bags," and "Smart Sewing with Cotton Bags."[37] The booklets also advertised Simplicity and McCall's patterns to help women make a variety of clothing. Each dress pattern in the booklet carried the same number as the regular McCall's and Simplicity patterns so women could order them directly from the companies. For these patterns, however, the companies translated the yardage of fabric needed for a dress into the number and size of the cotton bags required.

Many of the ideas suggested in the booklets borrowed from long-standing practices among rural women. The cotton council's 1949 booklet discussed "bag swappings." "Mrs. Jones brings two large cotton bags with rosebud print, hoping to find a neighbor with enough red checks to enable her to make a set of new kitchen curtains. Your home demonstration group, 4-H club, mother's club or church group might profit by similar 'swappings.'"[38] A 1950 cotton council booklet played on the theme of neighborliness to promote the use of cotton bags: "Your good works can be just old-fashioned neighborliness spread out. Things you can give others and do for others— things that can't be bought. You and your friends can find many new ways to use cotton bags in your community."[39]

The cotton, textile, and feed industries sponsored cotton-bag sewing contests and fashion shows to promote sales of their products. The cotton council and individual mills stimulated interest in feed bag fashions by lending sample wardrobes to women's clubs. After Bewley Mills of Texas, manufacturers of chicken and livestock feed, hired a home economist to direct fashion shows, the company claimed that farm women started "pestering their men folk to lay in plenty of the Red Anchor Feed."[40] In 1950 the council's "Cotton Bag Loan Wardrobe" featured eighteen items made with McCall's patterns. The demand from women's groups kept twenty-four wardrobes in circulation.[41] Although sewing contests had been held locally during the 1930s, industry-sponsored competitions could reach a national audience. Contestants first entered their garments at county fairs, with winners advancing to the state, regional, and national levels. Finalists in the 1959 competition sponsored by the National Cotton Council and Textile Bag Manufacturers Association vied for an array of prizes, ranging from household appliances such as mixers and electric blankets to cash, a glittering vacation in Hollywood, and lunch at Sardi's in New York City.[42]

Women entered the sewing contests and ordered booklets by the thou-

Advertisement for cotton bag sewing contest, 1960. (Photo courtesy of Office of Imaging, Printing, and Photographic Services, Smithsonian Institution, Washington, D.C.)

sands. In 1945, the cotton council claimed that its booklet was in the hands of more than one million women; three years later the council boasted that it had received more than four thousand requests for its sewing booklet in one day and that its Cotton Bag Wardrobe Shows had attracted more than thirty thousand people in a four-month period.[43] A 1948 sewing contest sponsored by Farmers Cooperative Exchange stores in North Carolina attracted more than five hundred women.[44]

By 1960, however, the heyday of bag fashions had come and gone. Their demise coincided with changes in the farm economy and the bag industry. Paper bags edged out cloth packaging, and as livestock and poultry operations expanded, dealers began delivering feed in bulk and dumping it into large grain bins. And bulk feed eliminated a source of income for women like

Arthur Fleming's wife and her neighbors. "Oh, they didn't like it," he recalled, "but there wasn't nothing you could do about it, see."[45]

VIEWED IN NEW contexts, the prosaic feed bag can assume surprisingly complex meanings. While sewing with sacks has come to represent "making do" during hard times in the popular imagination and collective memory, by some measures the availability of bags for reuse was as much a sign of prosperity as a sign of poverty. At the very least, they marked changes in the southern farm economy that had ambiguous consequences for women. Like women who negotiated with itinerant merchants or who took what they had and turned it into money, women who used and swapped feed sacks were linked to marts of trade that reached far beyond their neighborhoods. A woman who incorporated feed bags into a waste-not, want-not philosophy was also enmeshed in big businesses and marketing strategies that redefined sacks as a sign of fashion and modernity. Older values blended with new economic realities. Today a symbol of simpler times on the farm, feed bags also represent economic and cultural changes more complex than historians have imagined or than participants themselves often realized.

Introduction

1. See Evans, *Born for Liberty*, 203–4; Woloch, *Women and the American Experience*, 285–86; and Kerber and Mathews, *Women's America*, 425–31. All of these authors rely upon Hagood, *Mothers of the South*, for their brief descriptions of farm women in the 1930s.

2. On the "New Englandization" of American women's history, see Clinton, *Half-Sisters of History*, 1–17; and Hall, "Partial Truths." In a review essay on southern women's history published more than a decade ago, Jacquelyn Dowd Hall and Anne Firor Scott called for more attention to rural women; see Hall and Scott, "Women in the South," 477–80. Two scholars who have answered the call with subregional studies are Sharpless, *Fertile Ground, Narrow Choices*, and Walker, *All We Knew Was to Farm*.

3. In 1940, 63 percent of the South's population was rural, while the percentages of rural population in the North and West were 33 and 41 percent, respectively. Of the South's rural population, 39.2 percent lived on farms. The percentage of rural-farm population of selected southern states include N.C. (46.4), S.C. (48.1), Ga. (43.7), Tenn. (43.6), Miss. (64.1), and Ark. (57.0). See U.S. Department of Commerce, Bureau of the Census, *Sixteenth Census of the United States: Population*, 51.

4. Pete Daniel, curator of the Division of the History of Technology, was co-investigator. For fuller descriptions of this project, see Jones and Osterud, "Breaking New Ground"; Osterud and Jones, "'If I Must Say So Myself'"; Jones "'Mama Learned Us to Work'"; and Jones, "Voices of Southern Agricultural History."

5. This overview is drawn from such studies as Hahn, *Roots of Southern Populism*; Daniel, *Breaking the Land*; Fite, *Cotton Fields No More*; Kirby, *Rural Worlds Lost*; and Wright, *Old South, New South*.

6. On oral history as a collaborative process, see Frisch, *A Shared Authority*. The pioneer of rural women's history was Joan M. Jensen. Her anthology, *With These Hands*, set a research agenda and guided scholars to the variety of sources available. Jensen also organ-

ized the First Conference on the History of Rural/Farm Women in Las Cruces, N.M., in 1984. Papers read at the first and subsequent conferences previewed important articles, anthologies, and monographs that would follow. These include Jensen, *Loosening the Bonds*; Jensen, *Promise to the Land*; Fink, *Open Country, Iowa*; Fink, *Agrarian Women*; Osterud, *Bonds of Community*; Jellison, *Entitled to Power*; and Neth, *Preserving the Family Farm*.

7. Edward L. Ayers has cautioned, "It is dangerous to let southern poverty and oppression be the entire story of the South. Told often enough, exclusively enough, such stories unintentionally flatten southerners, black and white, into stock figures, into simple victims and villains." See "Narrating the New South," 564. Calculations derived from U.S. Department of Commerce, Bureau of the Census, *Fifteenth Census of the United States: Agriculture, the Southern States*, 30–31. For the evocative phrase "land-orphanage," see E. C. Branson quoted in Hobbs, *Know Your Own State—North Carolina*, 29. For studies that emphasize the successes and endurance of small landowning black and white farmers in the South, see Schultz, "The Dream Realized"; Schultz, "The Unsolid South"; and Petty, "Standing Their Ground."

8. Griffin interview, 5–8; Pender interview, 9; both in "An Oral History of Southern Agriculture," National Museum of American History, Smithsonian Institution, Washington, D.C. (hereafter cited as OHSA).

9. Murray interview, 6 January 1987, 12, 14, OHSA.

10. Ibid., 9.

11. Sharpless, *Fertile Ground, Narrow Choices*, 159–87; Allen, *Labor of Women in the Production of Cotton*, passim.

12. Daniel, *Breaking the Land*, 156–62.

13. Ibid., 23–31.

14. Winski interview, 15; Bennett interview, 10–11; Davis interview, 24–25; all in OHSA.

15. Pender interview, 17.

16. See Jones, "'Mama Learned Us to Work,'" which is a narrative braided from interviews conducted in Statesville, N.C., 4 May 1987 and 29 December 1987.

17. Murray interview, 9 January 1987, 29–33; Nellie Stancil Langley interview, 10, OHSA. For a brief account of cooking at tobacco barns, see Jeter "'Sitting Up' with the Fire Often Inspired Romance," 20.

18. Taylor, *Down a Country Road*, 60–62.

19. Dove interview, 9; Anderson interview, 15; and Purvis interview, 6; all in OHSA. Other descriptions of women as "good managers" on southern farms in the 1930s can be found in Terrill and Hirsch, *Such as Us*, 72–78, 99–102, 108–15. On the social construction of the role of "good manager," see May, "The 'Good Managers'"; and Benson, "Living on the Margin."

20. Odum, *Southern Regions*; Coclanis and Carlton, *Confronting Southern Poverty in the Great Depression*.

21. Definition of yeomanry from Edwards, *Scarlett Doesn't Live Here Anymore*, 33. Anne Firor Scott has noted the neglect of yeoman farm women in the nineteenth century and "the irony in the fact that we have more data about the poorest southern farm families, that is, tenant and share farmers, than we do about the independent landowners on various sizes of farms" in the twentieth century. See Scott, *Southern Lady*, 280–81, 286, 296–97 (n. 47). Similarly, Laura F. Edwards has observed that "common white women are the most invisible of all southern women in the post–Civil War South," and because of sparse documentation they have "slipped through the historical cracks." See *Scarlett Doesn't Live Here Anymore*, 150.

22. On "the federal road" to development in the South, see Kirby, *Rural Worlds Lost*, 51–79.

23. Here I am thinking about the implications for farm women of the Smith-Lever Act (1914), discussed below. Congress, of course, had passed protective legislation for wage-earning women before then. The Children's Bureau (1912) and the Women's Bureau (1920) of the U.S. Department of Labor studied conditions and made recommendations on behalf of rural and urban women and children. According to Molly Ladd-Taylor, farm women were among the most enthusiastic constituents of the Sheppard-Towner Act (1921) promoting maternal and infant care; Smith-Lever was its model. See Ladd-Taylor, "'My Work Came Out of Agony and Grief.'"

24. *Report of the Commission on Country Life*, quoted in Bowers, *Country Life*, 63–64; U.S. Department of Agriculture, Office of the Secretary, Report No. 103, *Social and Labor Needs of Farm Women*; U.S. Department of Agriculture, Office of the Secretary, Report No. 104, *Domestic Needs of Farm Women*; U.S. Department of Agriculture, Office of the Secretary, Report No. 106, *Economic Needs of Farm Women*.

25. Scott, *Reluctant Farmer*, 288–313; Baker, *County Agent*, 1–45.

26. Jones, "'The Task That Is Ours.'"

27. *Progressive Farmer*, 29 March 1913. See also Poe, *My First 80 Years*, esp. 87–106.

28. Campbell, *Movable School*, 85–86. On black agrarian progressives in Texas, see Reid, "Rural African Americans and Progressive Reform."

29. For accounts of these developments, see Leloudis, *Schooling the New South*; and Link, *Paradox of Southern Progressivism*.

30. Bailey, *Seaman A. Knapp*; Scott, *Reluctant Farmer*, 206–42; Young, "History of Agricultural Education in North Carolina."

31. For an overview of children's agricultural clubs, see Martin, *Demonstration Work*, 40–80.

32. Knapp's instructions to the Tomato Club agent can be found in Martin, *Demonstration Work*, 193.

33. On the Midwest, see Neth, *Preserving the Family Farm*, 130, 137–39; and Jellison, *Entitled to Power*, 53–54. On the Northeast, see Babbitt, "Productive Farm Woman." On the South, see Walker, *All We Knew Was to Farm*, chap. 4.

34. For a study that looks at rural women and reform in Canada from the "top down" and the "bottom up," see Ambrose and Kechnie, "Social Control or Social Feminism." A positive assessment of extension work among rural women in the Midwest and Far West is Holt, *Linoleum, Better Babies, and the Modern Farm Woman*.

35. McKimmon, *When We're Green We Grow*. Letter to *Alamance Gleaner* included in Edna Reinhart, Home Demonstration Annual Report, Alamance County, 1923, *Cooperative Extension Service Annual Reports*, reel 17, Microforms Collection, North Carolina State University Archives, D. H. Hill Library, Raleigh, N.C. (hereafter cited as *Annual Reports*). Only a portion of the letter was included; the name of the writer and the date of publication were not available. For examples of home demonstration clubwomen exercising political pressure, see "Club Work Given Credit by Leader: Mrs. Reid Cites Changes in Farm Life Brought by Home Demonstration Work," [1934], n.p., clipping in Scrapbooks, N.C. Federation of Home Demonstration Clubs, 1931–46; and clippings of letters to the editor of the *Smithfield Herald*, n.d., in Scrapbooks, Southeastern District, 1921–37; both in North Carolina State University Archives, D. H. Hill Library, Raleigh, N.C.; and Mrs. George E. Goforth to Gov. O. Max Gardner, 31 December 1932, "Agriculture and Home Economics, 1929–1933" folder, box 106, O. Max Gardner Papers, North Carolina Division of Archives and History, Raleigh, N.C. (hereafter cited as NCDAH).

36. Precise information about the class composition of North Carolina home demonstration members is hard to come by. According to a study conducted in the 1940s, 64 percent came from landowning families, 23 percent from tenant families, and the rest were wives and daughters of merchants and businessmen who serviced farm families. Results of this survey are reported in McKimmon, *When We're Green We Grow*, 55. On home demonstration among African American women in other southern states, see Reiff, "'Rousing the People of the Land'"; and Harris, "Grace under Pressure."

37. On interior decorating and landscaping, see Jane S. McKimmon, "Report of the Division of Home Demonstration Work, 1924," p. 15, "1923–1924" folder, box 1, Jane S. McKimmon Papers, NCDAH; on making inexpensive hats and clothing, see Jane S. McKimmon, "Report of the Division of Home Demonstration Work, 1927–1928," pp. 49–50, "1928" folder, box 3, McKimmon Papers, NCDAH; on efficacy of inexpensive home improvements, see Annie M. Murray, Home Demonstration Annual Report, Guilford County, 1936, [n.p.], reel 73, *Annual Reports*; Lillian H. Andrews, Home Demonstration Annual Report, Bertie County, 1937, pp. 15–16, reel 78, *Annual Reports*. On purchases resulting from home demonstration–sponsored Housekeepers Week in twelve counties, see Jane S. McKimmon, "Report of the Division of Home Demonstration Work, December 1919–December 1920," pp. 11–12, "1920–1921" folder, box 1, McKimmon Papers, NCDAH.

38. Jane S. McKimmon, "Annual Report 1929 Home Demonstration Division," pp. 66–68, "1929" folder, box 4; Jane S. McKimmon, "Annual Report 1930 Home Demon-

stration Division," 55–60, "1930" folder, box 4; both in McKimmon Papers, NCDAH.

39. Knapp and Creswell, "Effect of Home Demonstration Work on the Community and the County in the South," 251.

40. The concept of "pastoralization" is from Boydston, *Home and Work*, 142–93.

41. Radway, *Reading the Romance*. On annual reports, see Jones, "Reading between the Lines." My thinking about the dialectical nature of home demonstration has been especially influenced by Ladd-Taylor, *Mother-Work*, 167–96; and Ladd-Taylor, "'My Work Came Out of Agony and Grief.'"

42. Clippings of letters from Mrs. O. L. Boyett and Mrs. L. G. Flowers to the editor of the *Smithfield Herald*, n.d., in Scrapbooks, Southeastern District, 1921–37, North Carolina State University Archives.

43. Clipping of letter from Miss Monevah Barbour to the editor of the *Smithfield Herald*, n.d., in Scrapbooks, Southeastern District, 1921–37, North Carolina State University Archives.

44. For membership figures, see Jane S. McKimmon, "Home Demonstration Clubs for Women [1916]," p. 9, in "1916–1917" folder, box 1, McKimmon Papers, NCDAH; Jane S. McKimmon, "Report of the Division of Home Demonstration Work, 1922," *Eighth Annual Report of the North Carolina Agricultural Extension Service*, p. 147, "1922–1923" folder, box 1, McKimmon Papers, NCDAH; Jane S. McKimmon, "Report of Home Demonstration Work for North Carolina 1932," p. 1, "1932" folder, box 5, McKimmon Papers, NCDAH; I. O. Schaub, "A Year of Definite Farm Progress [1935]," p. 37, reel 66, *Annual Reports*. When white home demonstration clubs joined the North Carolina Federation of Women's Clubs in 1924, its membership skyrocketed. In 1905, the federation had 550 members; in 1911, it had 2,300 members; in 1924, it claimed some 50,000 members. Cotten, *History of the North Carolina Federation of Women's Clubs, 1901–1925*, 195–96.

45. Trachtenberg, *Incorporation of America*.

46. An abridged version of chap. 13 of Hagood's *Mothers of the South*, entitled "Of the Tenant Child as Mother to the Woman," appears in Kerber and Mathews, *Women's America*, 425–31.

Chapter One

1. Crews, *A Childhood*, 73–74.

2. Ibid., 74–76.

3. The classic study remains Clark, *Pills, Petticoats, and Plows*. Recent studies include Ayers, *Promise of the New South*, chap. 4; Hale, *Making Whiteness*, 168–79; and Ownby, *American Dreams in Mississippi*, chaps. 1–4.

4. Excellent introductions to the literature are Pendergast, "Consuming Questions,"

and Stearns, "Stages of Consumerism." On the cultural history of consumerism, see Lears, *Fables of Abundance*. Studies sensitive to gender, class, region, and race include Benson, "Living on the Margin"; Cohen, "Encountering Mass Culture at the Grassroots"; Cohen, "Class Experience of Mass Consumption"; Barron, *Mixed Harvest*, chaps. 5–6; Schlereth, "Country Stores, County Fairs, and Mail-Order Catalogues"; Hanson, "Home Sweet Home"; and Weems, *Desegregating the Dollar*.

5. Determining the precise number of itinerant merchants is difficult, and even official numbers must be taken with a grain of salt. According to U.S. censuses, the number of peddlers and hucksters increased in several southern states between 1890 and 1930. For example, in 1890 and 1930 respectively, there were: in Alabama, 440 and 813 peddlers and hucksters; in Arkansas, 150 and 258; in Georgia, 689 and 1,114; in Kentucky, 835 and 857; in North Carolina, 287 and 343; and in Tennessee, 773 and 998. For figures, see U.S. Department of the Interior, Census Office, *Compendium of the Eleventh Census*, 402, 403, 411, 426, 434; and U.S. Department of Commerce, Bureau of the Census, *Fifteenth Census of the United States: 1930, Population*, 113, 154, 381, 591, 1204, 1518. Sales agents are listed separately, although changing census categories make comparisons over time difficult. In 1920, there were: in Alabama, 454; Arkansas, 329; Georgia, 966; Kentucky, 480; North Carolina, 374; South Carolina, 221; and Tennessee, 916. See Department of Commerce, Bureau of the Census, *Fourteenth Census of the United States*, 68, 69, 86, 105, 122. Scholars who have focused on the Northeast have argued that peddlers were obsolete by the later nineteenth century. See Chandler, *Visible Hand*, 217; Jaffee, "Peddlers of Progress," 533; and Lears, *Fables of Abundance*, 102. A study that finds itinerants thriving in the mid-nineteenth-century Midwest is Lyon-Jennings, "A Telling Tirade."

6. Scull, with Fuller, *From Peddlers to Merchant Princes*, 25–33; Chandler, *Visible Hand*, 56; Wright, *Hawkers and Walkers in Early America*; Lears, *Fables of Abundance*, 40–101, esp. 63–74; Jaffee, "Peddlers of Progress," esp. 528–31.

7. Atherton, "Itinerant Merchandising in the Antebellum South" (quotation on p. 54).

8. Milledgeville *Georgia Journal*, 11 July 1826, p. 4. Thanks to Joseph T. Rainer for providing this reference.

9. Atherton, "Itinerant Merchandising in the Antebellum South," 43, 47, 53–56; on retaliation against itinerants in the wake of John Brown's raid, see Mohr, *On the Threshold of Freedom*, 5; and Columbus (Ga.) *Daily Sun*, 5 December 1859, p. 3.

10. On the importance of viewing consumption from the perspective of buyers, see Benson, "Consuming Questions." For a study of peddlers in the antebellum South that analyzes the power dynamics of gender, race, and consumerism, see Rainer, "Yankee Peddlers and Antebellum Southern Consumers in the Arena of Exchange"; also Rainer, "The Honorable Fraternity of Moving Merchants."

11. For an example of the perpetuation of the trickster image, see Harris, *Folk Plays*, 200–228. An informal history of and tribute to peddlers in America, which observes that Jewish peddlers served African Americans and poor whites, is Golden, *Forgotten Pi-*

oneer, 41. Accounts of these murders are in Johnson, *Tales of Country Folks*, 48–49; Shelton, *Day the Black Rain Fell*, 7–12. For other accounts of murdered peddlers, see Atherton, "Itinerant Merchandising in the Antebellum South," 41–43; Adams, *All Anybody Ever Wanted of Me Was to Work*, 49; and Coltman, "A '90s Murder Mystery."

12. Grobman, "Peddler Black of Owen County," 4 (emphasis in original). On peddlers traveling a circuit, see Spears, "Where Have All the Peddlers Gone?" For another account of a familiar peddler known as "old George," see Watkins and Watkins, *Yesterday in the Hills*, 157–58.

13. Ayers, *Promise of the New South*, 81–82; Clark, *Pills, Petticoats, and Plows*; and Ownby, *American Dreams in Mississippi*, 82–97. As several historians have noted, the mail-order businesses of Sears, Roebuck and Company and Montgomery Ward & Company also brought manufactured goods right to the homes of rural southerners. Ordering from a catalog provided women and African Americans some autonomy and anonymity as consumers and allowed them to bypass local stores. Nonetheless, mail-order customers had to pay in cash and could not evaluate the quality of goods until they arrived—two shortcomings that buying from itinerants avoided. On mail-order buying in the South, see Ayers, *Promise of the New South*, 87–88; Hale, *Making Whiteness*, 176–78; and Ownby, *American Dreams in Mississippi*, 75–76. That rural southerners often dubbed the mail-order catalogs "wish books" suggests how close and yet how far away the goods they offered might be for people of humble means.

14. Ownby, *American Dreams in Mississippi*, 61–81 (quotations on p. 72); Hale, *Making Whiteness*, 168–79.

15. Clark, *Pills, Petticoats, and Plows*, 36; Faulkner, *Light in August*, 20, 23. Pictures of country stores by Farm Security Administration photographers include Dorothea Lange, "A country store located on a dirt road, on Sunday afternoon," LC-USF-34-19911-C, Gordonton, N.C., July 1939; Jack Delano, "The interior of a general store," LC-USF-34-40652-D, Stem, N.C., May 1940; Delano, "The interior of the general store. High school boys on holiday because it is election day," LC-USF-34-40575-D, Stem, N.C., May 1940; Delano, "The interior of a general store," LC-USF-34-40652-D, Stem, N.C., May 1940; Delano, "Mr. Jackson in his general store," LC-USF-34-44483-D, Siloam, Ga., June 1941; all in Farm Security Administration/Office of War Information Collection, Division of Prints and Photographs, Library of Congress (hereafter cited as FSA/OWI). Ted Ownby explores poor and middling rural men as the family shoppers in antebellum general stores and notes that the custom was still intact in the 1930s; see *American Dreams in Mississippi*, 7–32, 102. To be sure, general stores were not entirely foreign to women, but poster-size advertisements featuring "cheesecake" images may have also added to female patrons' discomfort; see Russell Lee, "A Negro woman trading a sack of pecans for groceries in the general store," LC-USF-34-31759-D, Jareau, La., October 1938; and Russell Lee, "A sign in a country store," LC-USF-33-11805-M5, Vacherie, La., September 1938; both in FSA/OWI.

16. Simonsen, *You May Plow Here*, 56–58.

17. Ibid., 57; Griffin interviews, 108–9, "An Oral History of Southern Agriculture," National Museum of American History, Smithsonian Institution, Washington, D.C. (hereafter cited as OHSA).

18. Griffin interviews, 109.

19. Harris, *Folk Plays*, xvii; Harris, *Southern Savory*, 19–26. For a thoughtful literary biography, see Yow, *Bernice Kelly Harris*.

20. Coleman, *Five Petticoats on Sunday*, vii, 23, 24.

21. Harris, *Southern Savory*, 21; Coleman, *Five Petticoats on Sunday*, 24, 25.

22. Mohamed interview, 9–10, OHSA.

23. Ibid., 10.

24. Edmonds interview, author's collection. See also Spears, "Rolling Store: A Reminiscence." John Lewis, civil rights activist and U.S. congressman from Georgia, fondly remembered the visits by the "Rolling Store Man" during his 1940s childhood in rural Alabama. See Lewis with D'Orso, *Walking with the Wind*, 33. For visual evidence of rolling stores, see Russell Lee, "The interior of a traveling grocery store which caters to people in rural regions," LC-USF-34-31340-D, Forrest City, Ark., September 1938; Lee, "A traveling grocery store," LC-USF-33-11521-M2, Forrest City, Ark., October 1938; Lee, "A Negro making purchases at a traveling grocery," LC-USF-33-11625-M4, Forrest City (vicinity), Ark., September 1938; Jack Delano, "Inside a rolling store showing merchandise for sale," LC-USF-34-44010-D, Heard County, Ga., April 1941; Delano, "The driver of a rolling store," LC-USF-34-44035-D, Heard County, Ga., April 1941; and Delano, "A rolling store," LC-USF-34-44009-D, Heard County, Ga., April 1941; all in FSA/OWI.

25. Edmonds interview.

26. Thompson interview, 15–17, OHSA.

27. Sarten interview, 31–32, OHSA.

28. Harris, *Sweet Beulah Land*, 6; Harris, *Folk Plays*, 201 (quotation). See also Harris, *Purslane*, 211–12. In *Red Hills and Cotton*, 115, Ben Robertson recalled that his grandmother patronized sales agents because she wanted "to encourage them to come again, for peddlers brought news, and my grandmother was interested in news of every kind and especially enjoyed gossip."

29. Harris, *Sweet Beulah Land*, 26; Harris, *Purslane*, 211, 212; Hagood, *Mothers of the South*, 133–36.

30. For mention of Rawleigh and Watkins salesmen, see Thompson interview, 16; Purvis interview, 93; and Gosney interview, 8; all in OHSA. On changes in manufacturing and marketing in the late nineteenth century, see Trachtenberg, *Incorporation of America*; and Strasser, *Satisfaction Guaranteed*. For the importance of taking class, ethnicity, race, and context into account when studying consumerism, see Cohen, "Encountering Mass Culture at the Grassroots."

31. *Rawleigh Industries*, 8–13; J. R. Watkins Medical Company, *Open Door to Success*, 1–20;

and *A Merry Christmas and a Happy New Year from the J. R. Watkins Co. and Your Watkins Dealer*. On Rawleigh's first wife, see Barrett, *History of Stephenson County*, 530–31; and obituary of Mrs. Minnie B. Rawleigh, Freeport *Journal Standard*, 19 November 1947. Locating information about these companies has required persistence, serendipity, and friendly assistance. I found Rawleigh and Watkins publications in places as diverse as the Library of Congress and Smithsonian Institution, Washington, D.C.; Baker Library, Harvard University, Cambridge, Mass.; and an antique store in Pittsboro, N.C. Thanks to Dorothy Glastetter, a volunteer at the Freeport, Ill., public library, and to Tena Barratt, reference librarian there, who provided materials on the Rawleigh Company and its founder, and to Marie Dorsch, archivist at the Winona County Historical Society, for supplying copies of vintage and recent publications from the Watkins Company, which still does business as Watkins Products, Inc.

32. [Rawleigh], *Guide Book to Help Rawleigh Retailers*, 23–26, 284; *Rawleigh's 1927 Good Health Guide, Almanac, Cookbook*; *Rawleigh's Stock and Poultry Raisers' Guide*, copy in folder 19, box 2, Dairy files, Warshaw Collection, Archives Center, National Museum of American History, Washington, D.C.; Watkins Company, *Open Door to Success*, 47–49.

33. *Rawleigh's Stock and Poultry Raisers' Guide*, 18, 19; Watkins Company, *Open Door to Success*, 16, 17, 28.

34. *Guide Book to Help Rawleigh Retailers*; *Rawleigh Methods*; Watkins Company, *Open Door to Success*.

35. Strasser, *Satisfaction Guaranteed*, 110–13 (quotation on 112); *Rawleigh Industries*, 6–13; *Rawleigh's Stock and Poultry Raisers' Guide*, 17–20; Watkins Company, *Open Door to Success*, 4–30, 45.

36. Watkins Company, *Open Door to Success*, 19; *Rawleigh's 1927 Good Health Guide*, 2.

37. [Rawleigh], *Guide Book to Help Rawleigh Retailers*, 42 (first quotation, emphasis in original), 47 (second and third quotations). On scientific salesmanship, see ibid., chap. 8. On the development of modern selling techniques and advice guides for commercial travelers, see Spears, *100 Years on the Road*, chap. 7.

38. For letters written—or purportedly written—by successful retailers, see [Rawleigh], *Guide Book to Help Rawleigh Retailers*, 115–214, esp. 129–31, 176, 195–96; and Watkins Company, *Open Door to Success*, 51–127.

39. [Rawleigh], *Guide Book to Help Rawleigh Retailers*, 119, 137–38, 144, 199.

40. Thomas interview, author's collection; [Rawleigh], *Guide Book to Help Rawleigh Retailers*, 88, 89, 253.

41. Thomas interview.

42. Jones interview, author's collection, 8 (second quotation); Edythe Hollowell Jones, letter to author, 1988 (in author's possession) (all other quotations).

43. Thomas interview; Harris, *Purslane*, 13, 90–91, 266; Jones, letter to author.

44. [Rawleigh], *Guide Book to Rawleigh Retailers*, 200; Jones, letter to author; Purvis interview, 93; Gosney interview, 8.

45. Russell Lee, "The family [caption obscured] and six children. They own twenty acres of ground on which there is a small, plainly furnished, but comfortable house," LC-USF-34-31787-D, New Iberia (vicinity), La., October 1938; Jack Delano, "Mrs. Fanny Parrott, the wife of an ex-slave," LC-USF-34-44162-D, Siloam (vicinity), Ga., May 1941; both in FSA/OWI.

46. Rawleigh, *Rawleigh Methods*, 1581–1625. The Watkins Company may have also directed particular attention to southern African Americans. Watkins Company, *Open Door to Success*, 39, lists two intriguing sales pamphlets—*Handling the Trade in the South* and *Handling the Watkins Line in Georgia*—but efforts to locate them have been unsuccessful. For an overview of race and consumerism, see Weems, *Desegregating the Dollar*.

47. Rawleigh, *Rawleigh Methods*, 1590 (first two quotations), 1583 (third quotation), 1594, 1608, 1623 (fourth quotation).

48. Ibid., 1603, 1612.

49. Ibid., 1595, 1599.

50. Booth interview, 13–14; Spivey interview, 34; both in OHSA. On the general practice of merchants limiting purchases by tenants, see Clark, *Pills, Petticoats, and Plows*, 55–58.

Chapter Two

1. Murray interview, 9 January 1987, 9–12, "An Oral History of Southern Agriculture," National Museum of American History, Smithsonian Institution, Washington, D.C. (hereafter cited as OHSA).

2. Murray interview, 6 January 1987, 44–45, OHSA.

3. Daniel, *Breaking the Land*; Kirby, *Rural Worlds Lost*; Fite, *Cotton Fields No More*.

4. On northeastern women's local trading, see Ulrich, *Good Wives*, 43–50; Ulrich, *A Midwife's Tale*, 72–101; and Boydston, *Home and Work*, 75–141. On midwestern women, see Faragher, *Sugar Creek*, 96–109, esp. 101–5. On premiums as advertising at the turn of the century, see Strasser, *Satisfaction Guaranteed*, 163–202. Thanks to Sarah Wilkerson-Freeman for drawing the link between the surplus soap wrappers and black women's wage work as laundresses.

5. McCurry, *Masters of Small Worlds*, 37–91.

6. Wood, *Women's Work, Men's Work*, 31–100, 185.

7. Holt, *Making Freedom Pay*.

8. A concise analysis of how women's production for market complemented commercial agriculture is Jensen, "Cloth, Butter, and Boarders: Women's Household Production for Market." Other useful sources include Sachs, "The Participation of Women and Girls in Market and Non-Market Activities on Pennsylvania Farms"; Flora and Stitz, "Female Subsistence Production and Commercial Farm Survival among Settlement Kansas Wheat Farmers"; and Hollingsworth and Tyyska, "The Hidden Producers." Southern farm families pursued a variety of strategies to reduce risk and to earn cash.

In poor families, in particular, adult men earned money by working at sawmills and taking other seasonal jobs off the farm. Black women often earned money by working as domestics or laundresses. Women's production for exchange was one among a limited number of options that farm families might consider as they sought to supplement their incomes. Jacqueline Jones has acknowledged the value of women's trade and described the various ways that poor farm families made ends meet in *Labor of Love, Labor of Sorrow*, 89–90, and *The Dispossessed*, 89–99, esp. 93.

9. Although scholars have established the importance of women's depression-era household production for market for farm families in the Midwest, students of southern agriculture have paid far less attention to this portion of the rural economy and remained focused on cash crops and the desperate poverty of the region's farmers. Promising correctives include Schultz, "The Unsolid South," chap. 1; and Pettey, "Standing Their Ground," chap. 4. On the Midwest, see Fink, "Sidelines and Moral Capital: Women on Nebraska Farms in the 1930s"; Fink and Schwieder, "Iowa Farm Women in the 1930s: A Reassessment."

10. Taylor and Zimmerman, "Economic and Social Conditions of North Carolina Farmers," 17–26.

11. Harris, *Folk Plays*, xii, 111–58.

12. Dillard interview, 19–20, OHSA; Jones, *For the Ancestors*, 24–25; Felknor interview, 8, OHSA.

13. Taylor and Zimmerman, "Economic and Social Conditions of North Carolina Farmers," 17–26.

14. Hamby, *Memoirs of Grassy Creek*, 175.

15. Taylor, *Sharecroppers*, 76; Langley interview, 17, 28–29; Foster interview, 11; both in OHSA.

16. Langley interview, 17; Dillard interview, 19–21; Hamby, *Memoirs of Grassy Creek*, 218–19.

17. Charles Alan Watkins, "The Ward General Store Exhibit," working paper typescript, n.d., Appalachian Cultural Museum, Appalachian State University, Boone, N.C. According to Strasser, *Satisfaction Guaranteed*, 73–74, a 1919 study conducted by the Harvard University Bureau of Research found that 87 percent of general merchandise stores reported receiving farm produce, especially butter and eggs, from customers and reimbursing them in cash, merchandise, or credit.

18. John Ward Store Collection, 1914 Day Book, pp. 6, 21, Appalachian Cultural Museum, Appalachian State University, Boone, N.C.

19. John Ward Store Collection, Invoice Ledger, 1911–12, pp. 21, 22.

20. Ibid., 27.

21. Ibid., 19.

22. Hamby, *Memoirs of Grassy Creek*, 184.

23. Edmonds interview, author's collection.

24. Rountree interview, 27–30, OHSA.

25. Ibid., 31, 46.

26. Ibid., 38–39.

27. Edmonds interview; Rountree interview, 34–35. For a critical assessment of hucksters as "nefarious dealers," see "Hucksters Have Tough Sledding in Carolina," 22.

28. Westling, *He Included Me*, 35; Margaret Christine Nelson interview by Mary Hebert, Summerton, S.C., 5 July 1995, "Behind the Veil: Documenting African-American Life in the Jim Crow South," tape 2, side A, Rare Book, Manuscript, and Special Collections Library, Duke University, Durham, N.C.; Player interview, 52, OHSA.

29. Murray interview, 6 January 1987, 44, 47–51, 53–54; Murray interview, 11 January 1987, 46–48, OHSA.

30. *Progressive Farmer*, 20 February 1915, p. 15 (hereafter cited as *PF*), and 8 April 1916, p. 6. On the use of the U.S. Postal Service for marketing farm products, see Flohr, "Marketing Farm Produce by Parcel Post."

31. Jane S. McKimmon, "Home Demonstration Work in North Carolina, December 1, 1921–December 1, 1922," p. 34, "1921–1926" folder, box 11, Home Economics Annual Reports, Agricultural Extension Service/Agriculture School Collection, North Carolina State University Archives, D. H. Hill Library, Raleigh, N.C.

32. On the decline in crop prices, see Tindall, *Emergence of the New South, 1913–1945*, 111–13. On curb markets in Alabama, see Flynt, *Poor but Proud*, 301–2; and Rieff, "'Rousing the People of the Land,'" 133–36. On markets in South Carolina, see *PF*, 30 April 1921, p. 13, and *PF*, April 1939, p. 58. On markets in Virginia, see McCleary, "Home Demonstration Club Movement"; and *PF*, February 1935, p. 30. For growth of North Carolina markets and sales figures, see Jane S. McKimmon, "Report of Home Demonstration Work, December 1, 1924–December 1, 1925," p. 89, "1924–1925" folder, box 3; Jane S. McKimmon, "Report of Home Demonstration Work for North Carolina 1932," p. 36, "1932" folder, box 5; Jane S. McKimmon, "Report of Home Demonstration Work, 1933," p. 92, "1933" folder, box 7; and Jane S. McKimmon, "Report of Home Demonstration Work, December 1, 1935–December 1, 1936," p. 32, "1935–1936" folder, box 3; all in McKimmon Papers, NCDAH.

33. Jane S. McKimmon, "Report of Division of Home Demonstration Work, 1922," pp. 118–22; Delano F. Wilson, Home Demonstration Annual Report, Mecklenburg County, 1926, pp. 4–7, *Cooperative Extension Service Annual Reports*, reel 29, Microforms Collection, North Carolina State University Archives, D. H. Hill Library, Raleigh, N.C. (hereafter cited as *Annual Reports*). In addition, because the titles of reports that local agents submitted each year are long and change frequently, I have chosen to identify them by the name of the agent, the county, the year, and the reel number of the microforms collection of the reports.

34. On markets as pedagogical sites, see description of the successful Rocky Mount,

N.C., market in *PF*, April 1935, p. 44; for the comments of the Carteret County agent, see Jane S. McKimmon, "Home Demonstration Annual Report, 1931," pp. 37–38, "1931" folder, box 5, McKimmon Papers, NCDAH.

35. For an example of rules and regulations, see Jane S. McKimmon, "Report of Home Demonstration Work, 1933," p. 109, McKimmon Papers, NCDAH. On the regulation of markets in England, see Schmiechen and Carls, *British Market Hall.*

36. Ann M. Mason, Home Demonstration Annual Report, Carteret County, 1933, p. 32, reel 58, *Annual Reports*; Violet Alexander, Home Demonstration Annual Report, Beaufort County, 1931, p. 146, reel 49, *Annual Reports*. In other times and places, market rules also encouraged decorum; on similar rules in Great Britain, see Schmiechen and Carls, *British Market Hall*, 170–72.

37. The racially exclusionary bylaws of the Pitt County, N.C., market are included in Jane S. McKimmon, "Report of Home Demonstration Work, 1933," p. 109, McKimmon Papers, NCDAH. For descriptions of the racially integrated city market in Raleigh, N.C., see Simmons-Henry and Edminston, eds., *Culture Town*, 65–67. For a discussion of African American women from the Sea Islands of South Carolina who sold at the Charleston city market, see Day, "Kinship in a Changing Economy: A View from the Sea Islands."

38. [Sarah J. Williams], Home Demonstration Annual Report, Columbus County, 1930, p. 14, reel 46, *Annual Reports*; Carrie S. Wilson, Home Demonstration Annual Report, Alamance County, 1932, p. 10, reel 57, *Annual Reports*.

39. Wilhelmina R. Laws, [Negro subject matter specialist], "1937 Annual Report," 23–24, reel 77, *Annual Reports*; Annie J. Johnson, Home Demonstration Annual Report, Rowan County, 1939, pp. 46–47, reel 100, *Annual Reports*.

40. Violet Alexander, Home Demonstration Annual Report, Beaufort County, 1931, p. 146, reel 49, *Annual Reports*; Virginia S. Sloan, Home Demonstration Annual Report, Carteret County, 1931, pp. 7–8, reel 49, *Annual Reports*; Jane S. McKimmon, "Home Demonstration Annual Report, 1931," pp. 41–42, 46, McKimmon Papers; NCDAH.

41. Cunningham interview, 33–35, OHSA.

42. Jane S. McKimmon, "Home Demonstration Annual Report, 1931," pp. 41–42, McKimmon Papers; NCDAH. Katherine Millsaps, Home Demonstration Annual Report, Edgecombe County, 1930, p. 2, reel 45, *Annual Reports*; Jane S. McKimmon, "Report of Home Demonstration Work, December 1, 1924–December 1, 1925," p. 96; Irma P. Wallace, Home Demonstration Annual Report, Cleveland County, 1931, p. 15, reel 49, *Annual Reports*; Mary Lee McAllister, Home Demonstration Annual Report, Cabarrus County, 1933, p. 35, reel 58, *Annual Reports*.

43. This comparison of marketing methods is suggested by U.S. Department of Agriculture, "Marketing Eggs," 5–9; and Benjamin and Pierce, *Marketing Poultry Products*, 224–27.

44. "Mrs. Foster Ricks," by Luline L. Mabry, folder #593, Federal Writers Project Papers, Southern Historical Collection, Louis Round Wilson Library, University of North Carolina, Chapel Hill, N.C; McKimmon, *When We're Green We Grow*, 150–91.

45. Anderson interview, 14–15, 19–20, OHSA; see also Cooley, *Homes of the Freed*, 103–4, 174.

46. Murray interview, 6 January 1987, 47.

47. Benton interview, 6–10; Harrington interview, 27–28; both in OHSA.

48. Davis interview, 12, 74; Byers interview, 13; Foster interview, 12, 16; all in OHSA. On the tension between production for use and production for exchange, see Weber, *Peasants into Frenchmen*, 134–35.

49. Cunningham interview, 35.

50. Jane S. McKimmon, "Annual Report of Home Demonstration Work, December 1, 1925–December 1, 1926," p. 117, "1926" folder, box 3, McKimmon Papers, NCDAH; M. Adna Edwards, Home Demonstration Annual Report, Buncombe County, 1930, p. 15, reel 45, *Annual Reports*; Anamerle Arant, Home Demonstration Annual Report, Alamance County, 1932, p. 24a, reel 53, *Annual Reports*; Jane S. McKimmon, "Report of Home Demonstration Work, 1933," p. 95, McKimmon Papers, NCDAH; Jane S. McKimmon, "Home Demonstration Annual Report, 1931," pp. 46–47, McKimmon Papers, NCDAH; Jane S. McKimmon, "Annual Report of Home Demonstration Work, December 1, 1927–December 1, 1928," pp. 133–34, "1928" folder, box 3, McKimmon Papers, NCDAH; Jane S. McKimmon, "Annual Report of Home Demonstration Work, December 1, 1925–December 1, 1926," p. 116, McKimmon Papers, NCDAH; Lillian M. Debnam, Home Demonstration Annual Report, Robeson County, 1935, p. 14, reel 69, *Annual Reports*.

51. U.S. Department of Commerce, Bureau of the Census, *Fifteenth Census of the United States: Agriculture, the Southern States*, 498, 435.

52. Felknor interview, 2 May 1987, 22–28, 42; Taylor, *Sharecroppers*, 134–36; Byers interview, 12; Young interview, 18–19, OHSA.

53. Mary W. Huffines, Home Demonstration Annual Report, Robeson County, 1936, pp. 41–43, reel 75, *Annual Reports*; Jane S. McKimmon, "Home Demonstration Annual Report, 1931," pp. 46–47, McKimmon Papers, NCDAH.

54. Jane S. McKimmon, "Home Demonstration Annual Report, 1931," pp. 46–47, McKimmon Papers, NCDAH; Jane S. McKimmon, "Report of Home Demonstration Work, 1933," pp. 105–6, McKimmon Papers, NCDAH; Rosalind Redfearn quoted in *PF*, September 1933, p. 5.

55. Turner interview, 37–38; Anderson interview, 18; Player interview, 52; Byers interview, 11; all in OHSA.

56. Sarten interview, 12–14, 31–33, OHSA.

57. Murray interview, 6 January 1987, 62.

58. Lena Hillard to Youngs Rubber Corporation, 10 February 1941, file 442, box 25,

Clarence J. Gamble Papers, Countway Medical Library, Boston, Mass.; quoted in Johanna Schoen, "'A Great Thing for Poor Folks,'" 70.

59. Murray interview, 6 January 1987, 49, 54–55; Murray interview, 11 January 1987, 46; Murray interview, 9 January 1987, 38, 63; Murray interview, 16 January 1987, 4; all in OHSA.

60. This line of reasoning is suggested by Riegelhaupt, "Saloio Women." Riegelhaupt argues that the Portuguese peasant women she studied who sold in a nearby town or worked as domestics there tapped into a broader network of information and contacts than did their husbands, whose work, day in and day out, kept them closer to the home village. The contacts that women made gave them access to patrons who could help in time of need and subtly shifted the balance of power in families and undermined patriarchal authority even though formal political and economic power still rested with men. For an example of a woman who solicited help and charity on behalf of poor neighbors from town women to whom she regularly sold butter, see Hagood, *Mothers of the South*, 211. On the social and political functions of markets, see also Bohannon and Dalton, *Markets in Africa*, 15–19.

Chapter Three

1. Patterson interview, 8, 14, "An Oral History of Southern Agriculture," National Museum of American History, Smithsonian Institution, Washington, D.C. (hereafter cited as OHSA). For another summary of Patterson's life, see "She's 95, but Still Feels Young, Active," *(Taylorsville, N.C.) Times Advantage*, 21 March 1987, pp. 1, 3. On the commercial production of pecans, see *Progressive Farmer* (hereafter cited as *PF*), 14 February 1920, p. 55.

2. Patterson interview, 8–14.

3. While the trade literature and other primary sources available for a scholarly study of poultry are voluminous, the standard survey of the industry remains what appears to be the work of an industry "insider" and promoter: Sawyer, *Agribusiness Poultry Industry*. Another overview that includes mythology, folklore, science, and economics is Smith and Daniel, *Chicken Book*.

Two examples of the telescoped timeframe are Kirby, *Rural Worlds Lost*, 355–60; and Fink, *Open Country, Iowa*, 135–60. In a case study of egg production in Iowa, Fink contends that by the 1950s agricultural educators and extension personnel had discredited women's care of chickens and had convinced farm men to specialize in poultry production rather than treating it as one part of the general farm operation. In this account, however, Fink underplays the fact that the size of laying flocks grew steadily during the 1930s and 1940s, at a time when women remained in charge of poultry production. In the county she studied, flocks averaged 146 hens in 1925; twenty years later the average flock had 202 hens. Flocks this size were far larger than the national average and would have produced hundreds of eggs. These Iowa farm women made poultry a substantial enterprise that must have required marketing outlets beyond local grocery stores. In other words, while the rhetoric of agriculture officials encouraged men to consider poultry a profitable commodity, women were already leading the way. For an analysis of the subject that invites international comparisons, see Bourke, "Women and Poultry in Ireland."

In *Country People in the New South*, 88–89, Jeannette Keith notes that in Tennessee's Upper Cumberland, poultry sales soared after railroads linked farms to markets at the turn of the twentieth century.

4. Jensen, "Butter Making and Economic Development in Mid-Atlantic America, 1750–1850" (quotation on p. 185); Jensen, *Loosening the Bonds*; and McMurry, *Transforming Rural Life*. Historians of dairying in England, Ireland, and Scandinavia have considered the gender implications of commercialization, too. See Valenze, "The Art of Women and the Business of Men"; Bourke, "Dairywomen and Affectionate Wives"; and Sommestad, "Able Dairymaids and Proficient Dairymen."

5. *Bulletin of the North Carolina State Board of Agriculture* 21, no. 2 (February 1900): 2–5, 25. For an earlier example of the promotion of poultry, see John Howell Cobb, "Address on Poultry by Hon. Howell Cobb, Delivered before the Georgia State Agricultural Convention, Held at Brunswick, Ga., February 12th, 1889," Rare Book Collection, Southern Pamphlet 6096, Wilson Library, University of North Carolina at Chapel Hill. Interestingly, Cobb did not mention the gender of poultry raisers. For remarkably similar developments abroad, see Bourke, "Women and Poultry in Ireland."

6. [No name given], County Agent Annual Report, Union County, 1917, *Cooperative Extension Service Annual Reports*, reel 4, Microforms Collection, North Carolina State University Archives, D. H. Library, Raleigh, N.C. (hereafter cited as *Annual Reports*); N. K. Rowell, County Agent Annual Report, Chowan County, 1933, p. 5, reel 58, *Annual Reports*; Jane S. McKimmon, "Report of Home Demonstration Work, December 1, 1924–December 1, 1925," p. 85, "1924–1925" folder, box 3, Jane Simpson McKimmon Papers, North Carolina Division of Archives and History, Raleigh, N.C. (hereafter cited as NCDAH). Because the titles of reports that local agents submitted each year are long and change frequently, I have chosen to identify them by the agent, county, year, and reel number of the microforms collection of the reports.

7. U.S. Bureau of the Census, *Twelfth Census of the United States*, ccxxiii–ccxxvii; U.S. Department of Commerce, Bureau of the Census, *Thirteenth Census of the United States*, 512.

8. "Poultry and Eggs," in U.S. Bureau of the Census, *Twelfth Census of the United States*, ccxxiii–ccxxxii.

9. White, "Home to Roost."

10. For an overview of these developments, see Sawyer, *Agribusiness Poultry Industry*, 23–28.

11. On college research, see ibid., 19–20. On poultry research in North Carolina, see Morris, *Poultry Can Grow at NCSU.*

12. *PF*, 4 March 1909, p. 6. For examples of Mrs. Deaton's columns, see *PF*, 13 May 1909, p. 14; *PF*, 10 June 1909, p. 14; and *PF*, 21 October 1909, p. 14.

13. See *PF*, 4 March 1909, p. 2.

14. *PF*, 4 March 1909, p. 4. According to *The American Heritage Dictionary of the English Language* (New York: American Heritage Publishing Co., 1970), 1227, water glass is the common name for sodium silicate, "Any of various water-soluble silicate glass compounds used as a preservative for eggs, in plaster and cement, and in various purification and refining processes."

15. *PF*, 26 September 1907, p. 7.

16. *PF*, 6 August 1908, p. 14. Another letter from McPherson appears in *PF*, 4 March 1909, p. 6.

17. *PF*, 30 January 1915, p. 6.

18. *PF*, 20 January 1911, p. 10.

19. *PF*, 26 October 1918.

20. *PF*, 4 April 1914, p. 17; for another example of a woman using the income from sales of eggs gathered on Sunday to pay her Missionary Society dues, see *PF*, 19 February 1921.

21. On women swapping settings of eggs, see Welborn interview, 12, OHSA; Mack Ivey Cline, "I Remember," June 1981, typescript, Rare Book, Manuscript, and Special Collections, William R. Perkins Library, Duke University, Durham, N.C.

22. Harris, *Purslane*, 213, 222. "Suck-egg" dogs inspired some of the earliest letters to *Progressive Farmer* from women in the late nineteenth century. On 21 February 1888, Mrs. F. S. Hogan of Raleigh overcame her deference and shyness about expressing an opinion "in the public prints" because she was so frustrated by a "good for nothing suck-egg dog" which roamed her and her husband's property and robbed her hens' nests. Mrs. Hogan's letter struck a responsive chord. On 13 March 1888, "A Farmer's Wife" urged legislators to pass a dog tax law that would keep disruptive canines under control and protect the interests of women who were trying to earn some money by raising chickens.

23. Langley interview, 24–28, OHSA.

24. *PF*, 11 March 1909, p. 18; *PF*, 28 January 1909, p. 16.

25. *PF*, 28 April 1909, p. 14.

26. *PF*, 13 May 1909, p. 14.

27. *PF*, 26 October 1918. Although the editor appended a comment that censured Meriam's practice of combining artificial and natural means of incubation as "a bad one," he did not censor the letter and dismiss her mixed methods altogether.

28. Sawyer, *Agribusiness Poultry Industry*, 32.

29. U.S. Department of Commerce, Bureau of the Census, *Thirteenth Census of the*

United States, 513; U.S. Department of Commerce, Bureau of the Census, *Fifteenth Census of the United States: 1930, Agriculture*, 685–90.

30. Chas. A. Sheffield, "The Production and Consumption of Agricultural Products in North Carolina," 3–4 December 1929, North Carolina Agricultural Extension Service of the North Carolina State College of Agriculture and Engineering, typescript in "State College–Extension Division–1929" folder, box 107, General Correspondence, 1929–1933, O. Max Gardner Papers, NCDAH. Despite such a phenomenal increase, North Carolina farms produced only 54 percent of the poultry and eggs consumed in the state.

31. *PF*, February 1933, p. 5. In the mid-1920s the Montgomery County, N.C., farm agent had no trouble accounting for the increased popularity of poultry in his section. Demand was up, and the prices for poultry products had not declined as much as those for other commodities after World War I. See Oliver R. Carrithers, County Agent Annual Report, Montgomery County, 1926, p. 9, reel 29, *Annual Reports*. According to one USDA study, during 1931 butter and poultry products were the only farm commodites to remain above pre–World War I prices. See Melvin, "Rural Life," 938.

32. *PF*, 20 January 1923, pp. 1, 10.

33. *Dixie Poultry Journal* (hereafter cited as *DPJ*), January 1928, p. 11.

34. *DPJ*, March 1927, pp. 12–13.

35. *DPJ*, January 1927, p. 10.

36. Ibid., pp. 11, 18.

37. *DPJ*, August 1927, pp. 5, 13.

38. Ibid., pp. 7, 15.

39. *DPJ*, December 1929, p. 18.

40. On Patterson selling to Bunch Hatchery, see "She's 95, but Still Feels Young, Active," *(Taylorsville, N.C.) Times Advantage*, 21 March 1987, p. 1. On Bunch Hatchery, see *DPJ*, February 1927, p. 27; *DPJ*, February 1930, pp. 6, 22 (quotation on p. 22); and *DPJ*, January 1936, p. 7.

41. A. R. Morrow, County Agent Annual Report, Iredell County, 1926, p. 10, reel 29, *Annual Reports*.

42. J. T. Monroe, County Agent Annual Report, Jones County, 1926, p. 16, reel 29, *Annual Reports*.

43. Earl [last name unclear], County Agent Annual Report, Madison County, 1926, p. 11, reel 29, *Annual Reports*.

44. Addie Houston, Home Demonstration Annual Report, Guilford County, 1930, p. 14, reel 46, *Annual Reports*.

45. *DPJ*, October 1927, p. 9.

46. *Poultry Industry of the United States of America*, 34.

47. Slocum, "Marketing Eggs"; Lewis and James, "Poultry Products Marketing Survey"; Benjamin and Pierce, *Marketing Poultry Products*, 222–31.

48. *DPJ*, February 1927, pp. 7, 19.

49. Jane S. McKimmon, "Report of Division of Home Demonstration Work, 1922," *Eighth Annual Report, North Carolina Agricultural Extension Service*, pp. 121–22, "1922–1923" folder, box 1, McKimmon Papers, NCDAH.

50. Jane S. McKimmon, "Annual Report of Home Demonstration, December 1, 1922–December 1, 1923," p. 121, "1922–1923" folder, box 1, McKimmon Papers, NCDAH.

51. Lewis, Risher, and Salter, "Carlot Marketing of Poultry in North Carolina," 9; see also extension service annual reports for examples of car-lot marketing.

52. Jane S. McKimmon, "Report of Home Demonstration Work, December 1, 1924–December 1, 1925," pp. 102–3, McKimmon Papers, NCDAH.

53. E. P. Welch, County Agent Annual Report, Beaufort County, 1931, p. 11, reel 49, *Annual Reports*.

54. R. W. Pou, County Agent Annual Report, Forsyth County, 1931, p. 21, reel 50, *Annual Reports*.

55. Lillie H. Hester, Home Demonstration Annual Report, Bladen County, 1932, p. 11, reel 53, *Annual Reports*.

56. Ibid., p. 14, reel 62.

57. Hugh Overstreet, County Agent Annual Report, Carteret County, 1931, p. 11, reel 49, *Annual Reports*.

58. W. I. Smith, County Agent Annual Report, Durham County, 1931, p. 57, reel 49, *Annual Reports*.

59. *PF*, February 1933, p. 5.

60. *PF*, June 1935, p. 7.

61. This is in contrast to developments in dairying, as described in sources cited in note 6 above.

62. Murray interview, 6 January 1987, 41, 43, 51–53, 78, OHSA.

63. *PF*, 3 April 1926, p. 12.

64. Fleming interview, 14–18, OHSA.

65. Ibid., 1–2, 8–9. Years later, the pioneer home demonstration agent in Hall County, Ga., where Gainesville is located, claimed to have introduced the first eggs that launched the poultry industry in the area. See "Miss Blanch Welchell Returns to County," *Daily Times*, 6 June 1961, p. 5.

66. Fleming interview, 14–18; Byers interview, 16–17; both in OHSA.

67. Roberts interview, 20–25, OHSA.

68. I. O. Schaub, "Report of the Agricultural Extension Service of North Carolina State College for the year ending December 1, 1937," p. 42, reel 77, *Annual Reports*.

69. C. F. Parrish, "Report of Poultry Extension Work in North Carolina, December 1, 1937 to November 30, 1938," p. 2, reel 83, *Annual Reports*.

70. C. F. Parrish, "Report of Poultry Extension Work in North Carolina, 1940," p. 3, reel 103, *Annual Reports*.

71. Lovette paraphrased in Arndt, "Locational Considerations in the North Carolina Broiler Industry," 88. For an overview of the South, see Lord, "The Growth and Localization of the United States Broiler Chicken Industry."

72. Kilby interview, 10–12, OHSA; "Small Farm Route in Wilkes Has Become Million-Dollar Industry," *Winston-Salem Journal and Sentinel*, 6 November 1955, p. 9C; videotaped interview with Mrs. Charlie Lovette, Wilkes County Telecommunications Project, #892, copy 1, Wilkes Community College Library, North Wilkesboro, N.C., [no date; interviewer not identified].

73. Mrs. Charlie Lovette interview; "The Impact of the C. O. Lovette Family on the Poultry Industry," typescript, vertical files, Wilkes Community College Library, North Wilkesboro, N.C., [no date; no author.]

74. On the Lovette family, see Absher, *Heritage of Wilkes County*, 319; Simpson, *Heritage of Wilkes County*, 329.

75. For an example of the rhetoric and imagery of rural manliness during the war, see *PF*, May 1943, pp. 9, 26. Sawyer, *Agribusiness Poultry Industry*, 72–84.

76. C. F. Parrish, "Report of Poultry Extension Work in North Carolina, 1943," p. 1, reel 129, *Annual Reports*.

77. On men's reliance upon women and children, see Fleming interview, 34; and Byers interview, 18–21. See also W. H. Miller, "Fleming Ready for Competition," *Broiler Growing*, February 1951, pp. 14, 36–37, on Arthur Fleming's dependence upon family labor; and various issues of *Broiler Growing* in the 1950s for similar patterns.

78. One of the best critical examinations of contract broiler growing in the South is Shand, "Billions of Chickens."

Chapter Four

1. Charlotte Hilton Green, "Daughters of the New South," *Forecast*, June 1922, clipping in Conceit Book, 1915–1930, box 23, Jane Simpson McKimmon Papers, North Carolina Division of Archives and History, Raleigh, N.C. (hereafter cited as NCDAH).

2. Anne Pauline Smith and Frank O. Alford Papers, 1904–67, NCDAH. I am indebted to George Stevenson, personal manuscripts archivist, for alerting me to this valuable collection. My thinking about home agents' professional culture and identity has been shaped by Garrison, *Apostles of Culture*; Melosh, *"The Physician's Hand"*; Muncy, *Creating a Female Dominion*; Walkowitz, "The Making of a Feminine Professional Identity: Social Workers in the 1920s"; and Walkowitz, *Working with Class*. On changing gender roles, see Filene, *Him/Her/Self*, chaps. 1–5.

3. Useful introductions to the work of home demonstration agents include Ellenberg, "'May the Club Work Go on Forever'"; Hoffschwelle, *Rebuilding the Rural Southern Community*, chap. 5; McCleary, "Home Demonstration and Domestic Reform in Rural Vir-

ginia, 1910–1940"; Reiff, "'Rousing the People of the Land'"; and Rieff, "'Go Ahead and Do All You Can.'"

4. Jones, "'The Task That Is Ours'"; Scott, *Reluctant Farmer.*

5. Jones, "'The Task That Is Ours,'" chap. 4.

6. McKimmon, *When We're Green We Grow*, 14–15; Jane S. McKimmon, "Annual Report—1912–1913, the North Carolina Canning Club for Girls," "1912–1913" folder, box 1, McKimmon Papers, NCDAH. Among the married agents was Mrs. Rosalind Redfearn of Anson County; see "National Recognition Is Given to Mrs. Redfearn," *Wadesboro (N.C.) Messenger and Intelligencer*, 21 November 1935.

7. Smith and Wilson, *North Carolina Women*, 250–53; Rogers, *Tar Heel Women*, 279–82. Rogers speculated that if North Carolinians were asked to name "the woman who has contributed the most toward the enrichment of the lives of Tar Heel women," McKimmon would win the poll. For another glowing contemporary assessment of McKimmon, see Nell Battle Lewis, "Miss Jinnie and the Blossoming Desert," *Raleigh News and Observer*, 14 February 1937, clipping in Conceit Book—1915–1930, box 32, McKimmon Papers, NCDAH. More recently McKimmon was named among the most outstanding graduates of North Carolina State University; see Rob Christensen, "The Unofficial Top Ten List of 20th-Century NCSU Graduates," *Raleigh News and Observer*, 29 September 1997, p. 3A.

8. McKimmon, *When We're Green We Grow*, 1–19, esp. 9–10. Like many female reformers and professionals of her generation, in her memoir McKimmon downplayed the active role she played in shaping her career; on this phenomenon, see Conway, *When Memory Speaks*, 48–56.

9. On Smith's mother, see Pauline Smith to Frank Alford, May 6, 1939, "Private Correspondence, 1939" folder, box 3; on Smith's early work history, see Pauline Smith to Frank Alford, 11 January 1936, "Private Correspondence, 1935–1936" folder, box 3; both in Smith-Alford Papers, NCDAH. For biographical sketch, see George Stevenson, "Preliminary Description of Smith-Alford Papers," NCDAH.

10. Sims, *Power of Femininity*, chap. 3; Leloudis, *Schooling the New South*, 73–106, 155–73; Scott, *Southern Lady*, chaps. 5, 6. On American women's voluntary associations during the Progressive Era, see Scott, *Natural Allies*, chaps. 5, 6.

11. Ruth Evans Dozier to Catherine D. Thompson, 1 November 1974, Anne Firor Scott Collection, Rare Book, Manuscript, and Special Collections, William R. Perkins Library, Duke University, Durham, N.C. Dozier was ninety years old when she wrote the letter to Thompson, one of Scott's students.

12. Ruth Evans Dozier to Catherine D. Thompson, 1 November 1974, Scott Collection.

13. Ibid. Dozier's description of her encounter with unfamiliar conditions echoes the experiences of middle-class charity workers in turn-of-the-twentieth-century England described in Ross, *Love and Toil*, 11–26.

14. Reeder, *The People and the Profession*; McKimmon, *When We're Green We Grow*; Mary Wigley Harper, "The Wind Is from the East," 193–94, Collection 734, box 1, University Archives, Ralph Brown Draughon Library, Auburn University, Auburn, Ala. Reminiscences that illuminate the work of home agents in North Carolina can be found in the Mrs. Mamie Sue Jones Evans Papers, Johnston County Heritage Center, Smithfield Public Library, Smithfield, N.C. On the mobility of home demonstration agents, see Scharff, *Taking the Wheel*, 143–44.

15. McKimmon, *When We're Green We Grow*, 64–81.

16. On the public scrutiny of home agents, see *Business Opportunities for the Home Economist*, 248–51; and Harper, "The Wind Is from the East," 221–22, 231–32. On Margaret Martin's problems, see [D. L. Latham] to O. F. McCrary, 21 February 1921, D. L. Latham to Mrs. O. F. McCrary, 21 February 1921, Mamie Sue Jones to Jane S. McKimmon, 3 March 1921, Jane S. McKimmon to Mamie Sue Jones, 10 March 1921, Jane S. McKimmon to Mamie Sue Jones, 8 December 1921, and Mamie Sue Jones to I. E. Ketchum [Clerk of Board of Commissioners, Onslow County], 14 December 1921, "Letters of Mamie Sue Jones, 1921" folder, box 15, Smith-Alford Papers. For other examples where the individual performance of agents was considered a problem, see Mamie Sue Jones to Jane S. McKimmon, 28 January 1921, in ibid.

17. Maude Radford Warren, "She 'Lifted' Sixty-Six Counties: The Story of Jane S. McKimmon's Work in North Carolina," *Country Gentleman*, 29 June 1918, clipping in Conceit Book—1915–1930, box 32, McKimmon Papers, NCDAH. For a description of the desirable characteristics of home demonstration state leaders, see Nye, "The State Home Demonstration Leader and Her Program."

18. Ruth Evans Dozier to Catherine D. Thompson, 1 November 1974, Scott Collection; Jane S. McKimmon to Billy McKimmon, 3 January [1918], folder 2, box 1; Charles McKimmon to Billy McKimmon, 23 February 1918, folder 2, box 1; Billy McKimmon to Jane S. and Charles McKimmon, 26 August 1918, folder 3, box 1; Jane S. McKimmon to Billy McKimmon, 2 October 1918, folder 5, box 1; all in Jane Simpson McKimmon Collection, Southern Historical Collection, Louis Round Wilson Library, University of North Carolina at Chapel Hill (hereafter cited as SHC). On the apartment being rented in Jane McKimmon's name, see Jane S. McKimmon to Billy McKimmon, 4 April 1918, folder 3, box 1, ibid.

19. [Jane S. McKimmon], "Family News Sheet," 18 February 1918, folder 2, box 1, McKimmon Collection, SHC. For other examples of McKimmon's travels and speeches during the war, see [Jane S. McKimmon], "Family News Sheet," 26 February 1918, folder 2, box 1; and [Jane S. McKimmon], "Family News Sheet," 30 September 1918, folder 5, box 1, ibid. Charles McKimmon found part-time work in a bank in mid-1918; see [Jane S. McKimmon], "Family News Sheet," 8 September 1918, folder 5, box 1; and Jane S. McKimmon to Billy McKimmon, 21 February 1919, folder 7, box 1, ibid.

20. On her father's drinking, see Pauline Smith to Frank Alford, 13 August 1930,

"Private Correspondence, 1930 (June–December)" folder, box 2, Smith-Alford Papers, NCDAH. For more commentary on Smith's relationship to her father and his drinking habits, see Pauline Smith to Frank Alford, 23 August 1933, "Private Correspondence, 1933" folder, box 2; and Pauline Smith to Frank Alford, 10 August 1932, "Private Correspondence, 1932 (July–December)" folder, box 2, ibid. On her early work history, see Pauline Smith to Frank Alford, 11 January 1936, "Private Correspondence, 1935–1936" folder, box 3, ibid. In *Taking the University to the People*, 86–87, Wayne D. Rasmussen writes that "while teachers usually made higher salaries than home demonstration agents, prospects for moving up the ladder were better in Extension." Home agents also drew a salary year-round.

21. Stage, "Home Economics"; Weigley, "It Might Have Been Euthenics"; Rossiter, *Women Scientists in America*, 66–72; Cott, *Grounding of Modern Feminism*, 162–65.

22. Hoffschwelle, "The Science of Domesticity."

23. "Recommendations of the Committee on Extension Needs and Maintenance"; Willard, "Recruiting Extension Workers"; Frysinger, "The Home Economics Extension of the Future," 545.

24. Reiff, "'Rousing the People of the Land,'" 59–158; McKimmon, *When We're Green We Grow*; Martin, "Home Demonstration Work"; Frysinger, "The Home Economics Extension of the Future," 543–45; Lloyd, "Home Economics Extension—Purpose, Progress, and Prospect"; Sanderson, "Land-Grant Institutions and Rural Social Welfare"; *Business Opportunities for the Home Economist*, 246–53.

25. Calvin, "Extension Work"; Pritchett, "The Training of the County Agent"; Bunch, "A Course for Home Demonstration Agents"; *Business Opportunities for the Home Economist*, 248–49.

26. Pauline Smith to Frank Alford, [5 August 1929]; Pauline Smith to Frank Alford, [10 August 1929]; both in "Private Correspondence, 1929" folder, box 1, Smith-Alford Papers; Pauline Smith, Home Demonstration Annual Report, Franklin County, 1920, *Cooperative Extension Service Annual Reports*, reel 7, Microforms Collection, North Carolina State University Archives, D. H. Hill Library, Raleigh, N.C. (hereafter cited as *Annual Reports*). In addition, because the titles of reports that local agents submitted each year are long and change frequently, I have chosen to identify them by the name of the agent, the county, the year, and the reel number of the microforms collection of the reports.

27. Pauline Smith, Home Demonstration Annual Report, Franklin County, 1920, reel 7, *Annual Reports*.

28. Martin, *Demonstration Work*, 113.

29. For overviews of Smith's duties, see Smith, "Administrators Report, December 1, 1923–December 1, 1924," reel 19; Smith, "Administrative Report, 1924–1925," reel 23; Smith, "Administrative Report, 1925–1926," reel 27; "Administrative Report, 1928–1929," reel 39; "Administrative Report, 1930–1931," reel 48; "Administrative Report, 1932–1933," reel 57; all in *Annual Reports*.

30. Mamie Sue Jones to Jane S. McKimmon, 28 January 1921; Mamie Sue Jones to Jane S. McKimmon, 24 January 1921; both in "Letters of Mamie Sue Jones, 1921" folder, box 15, Smith-Alford Papers; Pauline Smith, "Administrators Reports, December 1, 1923–December 1, 1924," p. 2, reel 19, *Annual Reports*.

31. Pauline Smith, "Administrators Reports, December 1, 1923–December 1, 1924," p. 2, reel 19, *Annual Reports*.

32. Jane S. McKimmon, "Report of Division of Home Demonstration Work, December 1, 1924–December 1, 1925," pp. 5, 11, "1924–1925 folder," box 2, McKimmon Papers, NCDAH.

33. Ibid.; Jane Simpson McKimmon, B.S., M.S.," *N.C. State Alumni News* (December 1929): 1, clipping in Conceit Book #2, 1913–32, box 24, McKimmon Papers, NCDAH.

34. Martin, *Demonstration Work*, 96, 145–46. A regional field agent from the USDA echoed Martin's sentiments after visiting North Carolina and observing the work of agent Rosalind Redfearn in Anson County; see [Madge J. Reese] to Jane S. McKimmon, 13 December 1926, Conceit Book—1915–30, box 23, McKimmon Papers, NCDAH.

35. Pauline Smith to Frank Alford, 8 March 1930 and 31 May 1930, "Private Correspondence, 1930 (January–May)" folder, box 2, Smith-Alford Papers.

36. On the organizational structure of the extension service, see Baker, *County Agent*, chap. 5. A helpful introduction to issues of gender, organizational structure, and culture is Witz and Savage, "The Gender of Organizations."

37. Rachel Everett to Madge J. Reese, [November 1928]; Madge J. Reese to Rachel Everett, 27 November 1928; Rachel Everett to Madge J. Reese, 6 March 1929; Madge J. Reese to Rachel Everett, 11 March 1929; Rachel Everett to Florence E. Ward, 15 June 1929; Florence E. Ward to Rachel Everett, 25 June 1929; all in file "Misc. NC— 1928–1929," box 187, General Correspondence of the Extension Service and Its Predecessors, Records of the Extension Service, Record Group 33, National Archives and Records Administration, College Park, Md. (This series and record group will hereafter be referred to as Correspondence, RG 33.) Jane S. McKimmon, "Annual Report 1929 Home Demonstration Division," "1929" folder, box 4, McKimmon Papers, NCDAH. For another example of an agent who used home demonstration assignments to satisfy wanderlust, see Harper, "The Wind Is from the East," 217–45. On post–World War I opportunities in Europe, see Powell, "Home Demonstration Work in France."

38. Pauline Smith to Frank Alford, 6 December 1929, "Private Correspondence, 1929" folder, box 1; Pauline Smith to Frank Alford, [22 January 1930], "Private Correspondence 1930 (January–May)" folder, box 2, Smith-Alford Papers. On the politics of annual reports, see Jones, "Reading Between the Lines."

39. McKimmon, *When We're Green We Grow*, 65. Pauline Smith's various administrative reports reveal her talent for sizing up county politics.

40. Pauline Smith to Frank Alford, 22 July 1932, "Private Correspondence, 1932 (July–December)" folder, box 2; Pauline Smith to Frank Alford 12 May 1933, "Private

Correspondence, 1933" folder, box 2; both in Smith-Alford Papers. As evidence of their respect for her, clubwomen sometimes gave her holiday gifts; see Pauline Smith to "Dear Girls," 1 January 1946, "Professional Correspondence, 1946" folder, box 15, Smith-Alford Papers.

41. Mamie Sue Jones to Jane S. McKimmon, 24 January 1921, "Letters of Mamie Sue Jones, 1921" folder, box 15, Smith-Alford Papers; Jane S. McKimmon, "Report of Division of Home Demonstration Work, December 1, 1924–December 1, 1925," pp. 5, 11, McKimmon Papers, NCDAH; Harper, "The Wind Is from the East," 251–70.

42. Pauline Smith to Frank Alford, [Summer 1929]; Pauline Smith to Frank Alford, 5 August 1929; both in "Private Correspondence, 1929" folder, box 1, Smith-Alford Papers.

43. Pauline Smith to Frank Alford, November 1932, "Private Correspondence, 1932 (July–December)" folder, box 2, Smith-Alford Papers; Frank Alford to Pauline Smith, [June 1923], "Private Correspondence, 1923 (May–October)" folder, box 1, Smith-Alford Papers. On the culture of southern men, see Ownby, *Subduing Satan*. For insights into other romantic relationships, see unidentified man to Pauline Smith [n.d., circa 1910s], "Private Correspondence, 1904–1913" folder, box 1, Smith-Alford Papers; P. D. Stout to Pauline Smith, 27 August 1920, "Private Correspondence, 1920" folder, box 1; Smith-Alford Papers; Henry [T. Garris] to Pauline Smith, 4 March 1923, "Private Correspondence, 1923 (February–March)" folder, box 1, Smith-Alford Papers; see other letters from Garris to Smith in same folder.

44. Pauline Smith to Frank Alford, 17 April 1930; Pauline Smith to Frank Alford, 25 April 1930; Pauline Smith to Frank Alford, 2 May 1930; all in "Private Correspondence, 1930 (January–May) folder," box 2, Smith-Alford Papers.

45. Pauline Smith to Frank Alford, 26 March 1932, "Private Correspondence, 1932 (January–June)," box 2, Smith-Alford Papers. None of Smith's personal correspondence from 1931 survives in the collection. For summaries of her work, see Smith, "Administrative Report, 1930–1931," reel 48, *Annual Reports*, and Smith, "Administrative Report, 1932–1933," reel 57, *Annual Reports*.

46. "Action of County Commissioners in Cutting Budget Is Approved by Rising Vote of Citizens Monday," [Washington, N.C., newspaper], 5 July 1932, clipping in "Retrenchments, 1932–1933" folder, box 17, Smith-Alford Papers. By 1932, displays of support for home demonstration, which included testimonials from sympathetic dignitaries, district agents, and club members, were well rehearsed. For earlier examples, see Jane S. McKimmon to Mamie Sue Jones, 21 May 1921; Mamie Sue Jones to Jane S. McKimmon, 23 May 1921; Jane S. McKimmon to Mamie Sue Jones, 27 May 1921; and Judge Francis D. Winstead to Mamie Sue Jones, 25 June 1921; all in "Letters of Mamie Sue Jones, 1921" folder, box 15, Smith-Alford Papers. Jones was Smith's predecessor as district agent, and Smith retained these letters as part of the professional files that she inherited in 1922.

47. "Action of County Commissioners in Cutting Budget Is Approved by Rising Vote of Citizens Monday," [Washington, N.C., newspaper], 5 July 1932.

48. Pauline Smith to Frank Alford, 8 July 1932; Pauline Smith to Frank Alford, 12 July 1932; both in "Private Correspondence, 1932 (July–December)" folder, box 2, Smith-Alford Papers.

49. Pauline Smith to Frank Alford, 3 August 1932, "Private Correspondence, 1932 (July–December)" folder, box 2, Smith-Alford Papers; "Hertford County Slashing Its Budget," *Raleigh News and Observer*, 12 July 1932, p. 11.

50. Pauline Smith to Frank Alford, 3 August 1932, "Private Correspondence, 1932 (July–December)" folder, box 2, Smith-Alford Papers.

51. Pauline Smith to Frank Alford, 7 February 1933, "Private Correspondence, 1933" folder, box 2, Smith-Alford Papers; Smith, "Administrative Report, 1932–1933," p. 4, reel 57, *Annual Reports*.

52. Pauline Smith to Frank Alford, 12 July 1932, "Private Correspondence, 1932 (July–December)" folder, box 2, Smith-Alford Papers.

53. Pauline Smith to Frank Alford, 8 February 1932; Pauline Smith to Frank Alford, 21 March 1932; both in "Private Correspondence, 1932 (January–June)" folder, box 2, Smith-Alford Papers.

54. Pauline Smith to Frank Alford, 18 May 1933; Pauline Smith to Frank Alford, 9 October 1933; both in "Private Correspondence, 1933," folder, box 2, Smith-Alford Papers.

55. Pauline Smith to Frank Alford, 2 May 1930, "Private Correspondence, 1930 (January–May) folder"; Pauline Smith to Frank Alford, 1 March 1930, "Private Correspondence, 1930 (January–May) folder"; Pauline Smith to Frank Alford, 13 August 1930, "Private Correspondence, 1930 (June–December)" folder; all in box 2, Smith-Alford Papers.

56. Pauline Smith to Frank Alford, 13 August 1930, "Private Correspondence, 1930 (June–December)" folder, box 2, Smith-Alford Papers.

57. Pauline Smith to Frank Alford, 31 December 1936, "Private Correspondence, 1935–1936" folder, box 3, Smith-Alford Papers; Cookingham, "Combining Marriage, Motherhood, and Jobs before World War II"; Filene, *Him/Her/Self*, 130–37; Woloch, *Women and the American Experience*, 253–67. Although students of home demonstration in other southern states have found that before World War II female agents were expected to resign when they married, I have found no evidence of such a policy in North Carolina. In fact, according to McKimmon's successor, in 1937 four agents married and remained with the extension service; see Ruth Current, "Annual Report, 1937," p. 20, reel 77, *Annual Reports*. For expectations about married agents elsewhere, see Ellenberg, "'May the Club Work Go on Forever,'" 141–42; McCleary, "Home Demonstration and Domestic Reform in Rural Virginia," 45–46; and Reiff, "'Rousing the People of the Land,'" 66–67, 166–67.

58. Pauline Smith to Frank Alford, 5 August 1932, "Private Correspondence, 1932 (July–December)" folder, box 2, Smith-Alford Papers.

59. Pauline Smith to Frank Alford, 2 August 1932, "Private Correspondence, 1932 (July–December)" folder, box 2, Smith-Alford Papers.

60. Pauline Smith to Frank Alford, 13 September 1932, "Private Correspondence, 1932 (July–December)" folder, box 2, Smith-Alford Papers.

61. Pauline Smith to Frank Alford, 11 January 1936, "Private Correspondence, 1935–1936" folder, box 3, Smith-Alford Papers.

62. Pauline Smith to Frank Alford, [n.d., probably 1930s], "Private Correspondence, [1935–1949] fragments," folder, box 4, Smith-Alford Papers.

63. Pauline Smith to Frank Alford, 22 September 1932, and Pauline Smith to Frank Alford, 17 August 1932, "Private Correspondence, 1932 (July–December)" folder, box 2; Pauline Smith to Frank Alford, 6 May 1930, "Private Correspondence, 1930 (January–May)" folder, box 2; all in Smith-Alford Papers.

64. On Alford's mother, see Pauline Smith to Frank Alford, [10 November 1932?], "Private Correspondence, 1932 (July–December)" folder; Pauline Smith to Frank Alford, 21 March 1932, "Private Correspondence, 1932 (January–June)" folder; Pauline Smith to Frank Alford, November 1932, "Private Correspondence, 1932 (July–December)" folder; all in box 2, Smith-Alford Papers.

65. Pauline Smith to Frank Alford, [12 February 1935?], "Private Correspondence, 1935–1936" folder, box 3, Smith-Alford Papers.

66. Pauline Smith to Frank Alford, 21 March 1932, "Private Correspondence, 1932 (January–June)" folder, box 2, Smith-Alford Papers.

67. Pauline Smith to Frank Alford, 28 December 1944, "Private Correspondence, 1944–1946," folder, box 3, Smith-Alford Papers.

68. Pauline Smith to Frank Alford, 9 October 1933, "Private Correspondence, 1933" folder, box 2, Smith-Alford Papers.

69. Pauline Smith to Frank Alford, 9 January 1932, "Private Correspondence, 1932 (January–June)" folder, box 2, Smith-Alford Papers.

70. Pauline Smith to Frank Alford, 10 August 1932, "Private Correspondence, 1932 (July–December)" folder, box 2, Smith-Alford Papers.

71. Pauline Smith to Frank Alford, 4 January 1932, "Private Correspondence, 1932 (January–June)" folder, box 2, Smith-Alford Papers.

72. Pauline Smith to Frank Alford, 13 September 1932, "Private Correspondence, 1932 (July–December)" folder, box 2, Smith-Alford Papers.

73. Pauline Smith to Frank Alford, 23 September 1936, "Private Correspondence, 1935–1936" folder, box 3, Smith-Alford Papers.

74. Pauline Smith to Frank Alford, 23 June 1936, "Private Correspondence, 1935–1936" folder, box 3, Smith-Alford Papers.

75. Pauline Smith to Frank Alford, 22 September 1936, "Private Correspondence, 1935–1936" folder; Pauline Smith to Frank Alford, 27 December 1937, "Private Correspondence, 1935–1936" folder; both in box 3, Smith-Alford Papers.

76. Frank Alford to Pauline Smith, 15 January 1938, "Private Correspondence, 1937–1938" folder, box 3, Smith-Alford Papers.

77. Frank Alford to Pauline Smith, 30 July 1938, "Private Correspondence, 1937–1938" folder, box 3, Smith-Alford Papers.

78. Dorothy Dix column, February 1930, clipping in "Newspaper Clippings—Relationships between the Sexes" folder, box 9, Smith-Alford Papers.

79. Walkowitz, "The Making of a Feminine Professional Identity," 1052; Filene, *Him/Her/Self*, 144–66.

80. Pauline Smith to Frank Alford, 29 May 1939, "Private Correspondence, 1937–1938" folder (this letter was apparently misfiled); Pauline Smith to Frank Alford, [12 February 1935?], "Private Correspondence, 1935–1936" folder; both in box 3, Smith-Alford Papers.

81. Pauline Smith to Frank Alford, 20 April 1939, "Private Correspondence, 1937–1938" folder, box 3, Smith-Alford Papers (this letter was apparently misfiled).

82. Pauline Smith to Miss Bertha Lee Furguson, 5 January 1938, "APS Professional Correspondence, 1926–1939" folder, box 15, Smith-Alford Papers.

83. "The Huddlers," 1931–41, notebook, box 16, Smith-Alford Papers. The minutes do not describe the contents of "Sense and Nonsense."

84. Jane S. McKimmon to Pauline Smith, 11 January 1928, "APS Professional Correspondence, 1926–1939" folder, box 15, Smith-Alford Papers.

85. Pauline Smith to Frank Alford, 26 November 1932, "Private Correspondence, 1932 (July–December)" folder, box 2, Smith-Alford Papers.

86. Pauline Smith to Frank Alford, [January–February 1935?], "Private Correspondence, 1935–1936" folder, box 3, Smith-Alford Papers.

87. Virginia Edwards to Pauline Smith, 9 April 1939, "APS Professional Correspondence, 1926–1939" folder, box 15, Smith-Alford Papers.

88. Pauline Smith to Frank Alford, 20 April 1939, "Private Correspondence, 1937–1938" folder, box 3, Smith-Alford Papers (this letter was apparently misfiled).

89. Mary T. Knight to Pauline Smith, 1 January 1941, "Private Correspondence, 1940–1943" folder, box 3, Smith-Alford Papers.

90. Mrs. John M. Glenn to Pauline Smith, 30 December 1946, "APS Professional Correspondence, 1946" folder, box 15, Smith-Alford Papers.

91. Eloise B. Perry to Pauline Smith, 5 July 1941, "APS Professional Correspondence, 1940–1941" folder, box 15, Smith-Alford Papers. On professional women patterning management styles on quasi-kinship models, see Witz and Savage, "The Gender of Organizations," 42; Vicinus, "'One Life to Stand Beside Me'"; and Dean, "Covert Curriculum," 141–42.

92. Anne Harris to Pauline Smith, 9 March 1944, and Pauline Smith to Anne Harris, 13 May 1944, "APS Professional Correspondence, 1943–1945" folder, box 15, Smith-Alford Papers.

93. Iberria R. Tunnell to Pauline Smith, 24 February 1946, "APS Professional Correspondence, 1946" folder, box 15, Smith-Alford Papers.

94. T. L. McMullan to Pauline Smith, 6 February 1946, "APS Professional Correspondence, 1946" folder, box 15, Smith-Alford Papers.

95. For a summary of Smith's career, see the untitled typescript of what appears to be a press release announcing her retirement in 1949, "Miscellaneous—APS" folder, box 9, Smith-Alford Papers.

Chapter Five

1. Jane S. McKimmon, "Negro Women Praise Home Work," *Extension Farm News*, May 1923, clipping in scrapbook of newspaper clippings, 1923–1936, box 20, Jane S. McKimmon Papers, North Carolina Division of Archives and History, Raleigh, N.C. (hereafter cited as NCDAH). By the time McDougald began work in Wayne County, she was already an experienced community organizer. A native of Columbus County and a 1903 graduate of Elizabeth City State Normal School, McDougald had devoted her summer vacation of 1912 to securing pledges of $1,000 from African Americans in her home county to support the appointment of a black farm demonstration agent; her campaign was successful. See Brown, *E-Qual-ity Education in North Carolina among Negroes*, 91; and Brown, *A History of the Education of Negroes in North Carolina*, 125.

2. For ownership rate, see Hobbs, *North Carolina, Economic and Social*, 338; Emma L. McDougald, Home Demonstration Annual Report, Wayne County, 1924, *Cooperative Extension Service Annual Reports*, reel 16, Microforms Collection, North Carolina State University Archives, D. H. Hill Library, Raleigh, N.C. (hereafter cited as *Annual Reports*). In addition, the reports that local agents submitted each year will be identified by the name of the agent, the county, the year, and the reel number of the microforms collection of the reports.

3. Emma L. McDougald, Home Demonstration Annual Report, Wayne County, 1927, p. 12, reel 34, *Annual Reports*.

4. Articles that have emphasized the disabilities under which black agents worked include Hilton, "'Both in the Field, Each with a Plow'"; and Rieff, "'Go Ahead and Do All You Can,'" 134–52. Carmen Harris, on the other hand, has emphasized how black home agents initiated social change in "Grace under Pressure," and "'Fairy Godmothers' and 'Magicians.'"

5. Crosby, "Struggle for Existence"; Crosby, "Building the Country Home," chaps. 1, 2; Jones, "The South's First Black Farm Agents."

6. Seals, "The Formation of Agricultural and Rural Development Policy with Em-

phasis on African-Americans"; Crosby, "Struggle for Existence," 131; "Wilson Asked to Veto Bill," New York *Sun*, 5 May 1914, microfiche #6, Hampton University Newspaper Clippings File.

7. South Carolina's extension director balked at hiring black agents, arguing that if blacks agents were hired, white landowners would protest and would not allow them to work with their tenants for fear that they would sow seeds of discontent among their workers. White farmers, however, would not be as suspicious of white agents working with their tenants. See W. W. Long to J. A. Evans, 13 January 1915; W. W. Long to H. E. Savely, 13 January 1915; W. W. Long to Bradford Knapp, 29 January 1915; Bradford Knapp to W. W. Long, 13 February 1915; and W. W. Long to J. A. Evans, 15 February 1915; all in "South Carolina" file, box 7, General Correspondence of the Extension Service and Its Predecessors (1915), Records of the Extension Service, Record Group 33, National Archives and Records Administration, Washington, D.C. (hereafter cited as Correspondence, RG 33).

8. W. B. Mercier to C. R. Hudson, 14 January 1915; C. R. Hudson to W. B. Mercier, 20 January 1915; W. B. Mercier to C. R. Hudson, 23 January 1915; C. R. Hudson to Bradford Knapp, 26 February 1915; C. R. Hudson to Bradford Knapp, 22 April 1915; Bradford Knapp to C. R. Hudson, 23 April 1915; all in "North Carolina" file, box 7, Correspondence (1915), RG 33.

9. O. B. Martin to N. C. Newbold, 3 December 1913, box 1, General Correspondence of the Director, Division of Negro Education, Department of Public Instruction, NCDAH.

10. Gilmore, *Gender and Jim Crow*, 161–65, 197–98; Leloudis, *Schooling the New South*, 186–92, 201–4; Lasch-Quinn, *Black Neighbors*, 75–109. For quotation and the kinds of home economics and farm extension work under way at Hampton Institute in 1905, see Jones, *The Jeanes Teacher in the United States*, 16; and NASC Interim History Writing Committee, *The Jeanes Story*, 23.

11. Newbold, *Five North Carolina Negro Educators*, 61–85; Littlefield, "Annie Welthy Daughtry Holland."

12. Annie W. Holland to N. C. Newbold, 10 August 1914; Annie W. Holland to N. C. Newbold, 15 October 1914; both in box 3, Division of Negro Education, Department of Public Instruction, NCDAH.

13. Newbold, *Five North Carolina Negro Educators*, 61–85; Littlefield, "Annie Welthy Daughtry Holland," 569–70; B. T. McBryde to N. C. Newbold, 26 February 1917, box 3, Division of Negro Education, Department of Public Instruction, NCDAH; N. C. Newbold, *Circular Letter #1*, 3 March 1915, and *Circular Letter #2*, 3 July 1915; both in box 1, Division of Negro Education, Department of Public Instruction, NCDAH.

14. Jane S. McKimmon, "Report on Home Demonstration Work and Girls' Clubs," p. 38, "1917–1918" folder, box 1, McKimmon Papers, NCDAH.

15. Jane S. McKimmon, "Report on Home Demonstration Work and Girls' Clubs," p.

38; Jane S. McKimmon, "Summary of Home Demonstration Work in North Carolina, July 1, 1917 to July 1, 1918," pp. 48–51, box 1, McKimmon Papers, NCDAH; Gilmore, *Gender and Jim Crow*, 197–99.

16. Gilmore, *Gender and Jim Crow*, 197–99; Breen, "Southern Women in the War"; Jane S. McKimmon, "Report of Home Demonstration Work for Women and Girls, 1919," p. 22, "1919–1920" folder, box 1, McKimmon Papers, NCDAH.

17. Crow, Escott, and Hatley, *History of African Americans in North Carolina*, 132; Jane S. McKimmon, "Report of Home Demonstration Work for Women and Girls, 1919," p. 2; Evans, "Extension Work among Negroes," 7–8.

18. Hahamovitch, *Fruits of Their Labor*, 86–112. "Negro Women Wanted to Pick Cotton Crop," *Memphis Commercial Appeal*, 11 August 1918; "'Work or Fight' for Negro Women Urged," *New York Tribune*, 22 September 1918; "Abuse of the 'Work or Fight' Slogan," *New York Age*, 28 September 1918; all clippings in microfiche #548, Hampton University Newspaper Clippings File.

19. C. H. Stanton to Bradford Knapp, 19 August 1918; Bradford Knapp to C. H. Stanton, 29 August 1918; both in file "Misc. N.C. 1918–1919," box 56, Correspondence (1918–19), RG 33.

20. Woodruff, "African-American Struggles for Citizenship."

21. Jane S. McKimmon, "Report of Home Demonstration Work for Women and Girls, 1919," p. 23.

22. Jane S. McKimmon, "Report of the Division of Home Demonstration Work, December 1919–December 1920," p. 27, "1920–1921" folder; Jane S. McKimmon, "Report of Division of Home Demonstration Work in North Carolina, December 1920–December 1921," p. 6, "1921–1922" folder; both in box 1, McKimmon Papers, NCDAH.

23. Jane Simpson McKimmon, "Annual Narrative Report 1936 Home Demonstration Division," p. 14, "1936 Annual Report" folder, box 7, McKimmon Papers, NCDAH; Larkins, *Negro Population of North Carolina*, 20; Crow, Escott, and Hatley, *History of African Americans in North Carolina*, 125–26, 136–37; Brown, *A History of the Education of Negroes in North Carolina*, 123–31. On the important roles that black professionals played in building community institutions and pressing for government services, see Fairclough, "'Being in the Field of Education and Also Being a Negro . . . Seems . . . Tragic'"; Gavins, "Fear, Hope, and Struggle"; Hine, *Hine Sight*, 109–28; and Shaw, "*What a Woman Ought to Be and to Do*," esp. chap. 6.

24. Jane S. McKimmon, "Negro Women Praise Home Work"; Dazelle B. Foster, Home Demonstration Annual Report, Wake County, 1924, p. 3, reel 22, *Annual Reports*; Emma L. McDougald, Home Demonstration Annual Report, Wayne County, 1926, p. 3, reel 30, *Annual Reports*; and Emma L. McDougald, Home Demonstration Annual Report, Wayne County, 1925, p. 2, reel 26, *Annual Reports*.

25. Emma L. McDougald, Home Demonstration Annual Report, Wayne County, 1923, p. 1, reel 19, *Annual Reports*; Jane S. McKimmon, "Negro Women Praise Home Work."

26. Sanders, "Race Attitudes of County Officials." The unpublished report and comments collected from county officials are in the North Carolina Collection, Louis Round Wilson Library, University of North Carolina at Chapel Hill. According to an 11 June 1963 letter from Sanders to William Powell, curator of the North Carolina Collection, the professor withheld the material on race attitudes from publication for fear that it "might arouse misunderstanding and controversy." For more on the survey, see Crow, Escott, and Hatley, *History of African Americans in North Carolina*, 137–40.

27. All quotations come from unpaginated typed survey summaries that accompany Sanders, "Race Attitudes of County Officials." The survey summaries do not include the names of officials.

28. Ibid.

29. Ibid.

30. Ibid.

31. Gordon, "Black and White Visions of Welfare."

32. Jane S. McKimmon, "Annual Report, 1924," p. 122, "1924–1925 folder," box 1, McKimmon Papers, NCDAH; Dazelle B. Foster, Home Demonstration Annual Report, Wake County, 1924, p. 6, reel 22, *Annual Reports*; Dazelle B. Foster, Home Demonstration Annual Report, Wake County, 1925, pp. 4–5, reel 26, *Annual Reports*.

33. Dazelle Foster Lowe, "Report of Dazelle Foster Lowe, Negro District Agent, 1930," p. III-e, in Jane S. McKimmon, "Annual Report 1930 Home Demonstration Division," "1930" folder, box 4, McKimmon Papers, NCDAH. Efforts to find biographical information about Lowe have not been fruitful. According to the *Bulletin of the Agricultural & Technical College of North Carolina, Twenty-Ninth Annual Catalogue, 1933–1934, Vol. 25* (Greensboro, May 1934), 15, Lowe had earned a bachelor of science degree, but her alma mater was not given.

34. Emma L. McDougald, Home Demonstration Annual Report, Wayne County, 1924, reel 22, *Annual Reports*; Emma L. McDougald, Home Demonstration Annual Report, Wayne County, 1926, p. 24, reel 30, *Annual Reports*; Emma L. McDougald, Home Demonstration Annual Report, Wayne County, 1927, p. 14, reel 34, *Annual Reports*; Emma L. McDougald, Home Demonstration Annual Report, Wayne County, 1929, p. 19, reel 39, *Annual Reports*.

35. Higginbotham, *Righteous Discontent*, 185–229.

36. The ways in which black home demonstration agents used a conservative curriculum in subversive ways recalls the elasticity of the industrial education model; see Gilmore, *Gender and Jim Crow*, 160–65; Leloudis, *Schooling the New South*, 199–206; and Anderson, *Education of Blacks in the South*.

37. Hobbs, *North Carolina, Economic and Social*, 288–89; Woodson, *Rural Negro*, 4, 5, 8; Larkins, *Negro Population of North Carolina*, 32.

38. Smith, *Sick and Tired of Being Sick and Tired*, 1–74.

39. Sarah J. Williams, Home Demonstration Annual Report, Columbus County, 1931, p. 11, reel 49, *Annual Reports.*

40. Wilhelmina R. Laws, Home Demonstration Annual Report, Mecklenburg County, 1930, p. 12, reel 46, *Annual Reports;* Dazelle Foster Lowe, "1931 Annual Report for Negro Home Demonstration Agents," pp. 25–27, in Jane S. McKimmon, "Annual Report 1931 Home Demonstration Division," "1931" folder, box 5, McKimmon Papers, NCDAH; Emma L. McDougald, Home Demonstration Annual Report, Wayne County, 1930, p. 8, reel 48, *Annual Reports.*

41. Wilhelmina R. Laws, "1937 Annual Report Presented by Wilhelmina R. Laws, Negro Subject-Matter Specialist," pp. 27–28, reel 77, *Annual Reports;* Marietta Meares, Home Demonstration Annual Report, Craven County, 1937, p. 21, reel 78, *Annual Reports;* Lillian H. Andrews, Home Demonstration Annual Report, Bertie County, 1937, p. 21, reel 78, *Annual Reports.*

42. Jane S. McKimmon, "Report of Home Demonstration Work, 1933," pp. 1–12a, 48–52, "1933" folder, box 6, McKimmon Papers, NCDAH.

43. Dazelle Foster Lowe, "Eighth Annual Report of Extension Work as Conducted by the Negro Home Demonstration Agents of North Carolina, December 1, 1931–December 1, 1932," p. 35, in Jane S. McKimmon, "Report of Home Demonstration Work for North Carolina, 1932," "1932" folder, box 5, McKimmon Papers, NCDAH; Sarah J. Williams, "Narrative Report of Emergency Home Demonstration Work, Columbus County, May 1, 1933 to September 30, 1933," p. 4, "Home Economics Emergency Reports (Depression) 1933" folder, box 2, Director's Office Papers, Extension Home Economics, School of Agriculture and Life Sciences, North Carolina State University Archives, D. H. Hill Library, Raleigh, N.C.

44. "Narrative Report of Emergency Work, Johnston County," "Home Economics Emergency Reports (Depression) 1933" folder, box 2, Director's Office Papers, Extension Home Economics, School of Agriculture and Life Sciences; Mabel E. Oswalt, Emergency Home Demonstration Annual Report, Chatham County, 1934, p. 2, reel 62, *Annual Reports.*

45. Vanlandingham interview by Colvard, 1–5; Vanlandingham interview by the author, 1–8; both author's collection; Eugenia Patterson, "Narrative Report of Emergency Work, Washington County," [p. 5], "Home Economics Emergency Reports (Depression) 1933" folder, box 2, Director's Office Papers, Extension Home Economics, School of Agriculture and Life Sciences.

46. Jane S. McKimmon, "Annual Report 1934 Home Demonstration Division," p. 29, "1934" folder, box 6; Jane S. McKimmon, "Annual Narrative Report 1936 Home Demonstration Division," p. 87, "1936" folder, box 7; both in McKimmon Papers, NCDAH.

47. Dazelle Foster Lowe, "Tenth Annual Report of Dazelle Foster Lowe District

Home Demonstration Agent for Negroes 1934," p. 28, reel 62, *Annual Reports*, Jane S. McKimmon, "Annual Narrative Report 1935 Home Demonstration Division," pp. 16–17, "1935" folder, box 7, McKimmon Papers, NCDAH.

48. Dazelle Foster Lowe, "Tenth Annual Report," pp. 11–13, 27; Dazelle Foster Lowe, "Eleventh Annual Report, 1934–1935," pp. 2, 8, reel 67, *Annual Reports*.

49. Larkin, *Negro Population of North Carolina*, 19; Crow, Escott, and Hatley, *History of African Americans in North Carolina*, 140–41; Abrams, "Irony of Reform."

50. Studies of New Deal farm policies include Daniel, *Breaking the Land*; Kirby, *Rural Worlds Lost*; Conrad, *Forgotten Farmers*; and Grubbs, *Cry from the Cotton*.

51. For black protest against the AAA, see Wolters, *Negroes and the Great Depression*, 39–73; and the petition from Thomas M. Campbell, J. P. Davis, J. B. Pierce, and C. H. Waller to Cully A. Cobb, 7 March 1936, frame 96, reel 1, in Meier and Rudwick, *The Claude A. Barnett Papers*, pt. 3, *Subject Files on Black Americans, 1918–1967*, series A, *Agriculture, 1923–1966* (microform; hereafter cited as *Barnett Papers*).

52. Barnett had initially criticized New Deal farm programs but changed his tune in 1936; see Claude A. Barnett to Henry P. Fletcher, chair, Republican National Committee, 26 July 1934, frames 57–59, reel 1, *Barnett Papers*; Claude A. Barnett to Cully A. Cobb, 4 June 1935, frames 85–86, reel 1, *Barnett Papers*; and Claude A. Barnett to Cully A. Cobb, 7 September 1936, frame 109, reel 1, *Barnett Papers*.

53. Albon L. Holsey to Thomas M. Campbell, 6 November 1936, "Albon L. Holsey" folder, box 360; and Jennie B. Moton to Cully A. Cobb, 19 November 1936, "Mrs. Jennie B. Moton" folder, box 627; both in Central Correspondence Files, 1933–1947, Records of the Agricultural Adjustment Administration, RG 145, National Archives and Records Administration, College Park, Md. (hereafter cited as CCF, RG 145).

54. Although Jennie Booth Moton was married to one of the most prominent black leaders of the early twentieth century and participated in various groups, she has attracted little attention from historians. She did not merit an entry in the *Encyclopedia of Black Women in America*. According to Jacqueline Anne Rouse, fellow reformer Lugenia Burns Hope of Atlanta considered Moton an ineffectual spokeswoman for southern African American women; see Rouse, *Lugenia Burns Hope*, 112, 151–52 (n. 34). Margaret Murray Washington continued to head Women's Industries at Tuskegee and to assume other prerogatives of the "first lady" long after her husband's death in 1915. Thomas M. Campbell to E. A. Miller, assistant to the director, Southern Division, AAA, 12 October 1936, "T. M. Campbell" folder, box 114, CCF, RG 145.

55. Albon L. Holsey to Claude A. Barnett, 16 October 1936, frame 116, reel 1, *Barnett Papers*; Claude A. Barnett to Albon L. Holsey, 20 October 1936, frame 136, reel 1, *Barnett Papers*.

56. Dorothy Salem, "National Association of Colored Women," in *Black Women in America*; Hubbard interviews; Patterson interview; both in author's collection.

57. Moton Family Papers, boxes 18, 19, Manuscripts Division, Library of Congress,

Washington, D.C.; Jennie B. Moton to E. A. Miller, 15 March 1937, folder 19, box 18, Moton Family Papers.

58. J. P. Davis to Robert Russa Moton, 18 December 1936, folder 12, box 18, Moton Family Papers; Camilla Weems, Georgia assistant state agent for Negro work, to Jennie B. Moton, 17 February 1937, folder 6, box 19, Moton Family Papers.

59. P. H. Stone, Georgia state agent for Negro Work, to Jennie B. Moton, 15 February 1937, folder 5, box 19, Moton Family Papers; Helen M. Hewlette, Oklahoma district home demonstration agent, to Jennie B. Moton, 22 May 1937, folder 16, box 18, Moton Family Papers; Cassa L. Hamilton, Arkansas district home demonstration agent, to Jennie B. Moton, 10 November 1936, folder 18, box 18, Moton Family Papers.

60. Jennie B. Moton to Helen M. Hewlett, 22 July 1938, folder 16, box 18, Moton Family Papers.

61. "Report of Conditions among Negro Farmers of the Southern Region Agricultural Adjustment Administration," frames 482–97, reel 1, *Barnett Papers*.

62. Chas. A. Sheffield, Field Agent, Southern States, to Mrs. Elizabeth L. Brown, Edenton, N.C., 6 October 1939; Sheffield to William Dunn, Wendell, N.C., 16 November 1939; C. J. Ford, Roxboro, N.C., to C. W. Warburton, Director of Extension Work, Washington, D.C., 17 January 1940; W. H. Conway, Associate Chief, Division of Business Administration, to C. J. Ford, Roxboro, N.C., 30 January 1940; all in "North Carolina-A–F," folder, box 690, Correspondence, RG 33.

See also E. H. Shinn, Senior Agriculturalist, Surveys and Reports Section, to Thomas E. Moultrie, Dunn, N.C., 19 December 1939; Chas. A. Sheffield to Raymond Perry, Wendell, N.C., 21 December 1939; Sheffield to Carl Atvie Perry, Wendell, N.C., 16 November 1939; Mott Redfearn, Marshville, N.C., to U.S. Department of Agriculture, [December 1939]; E. H. Shinn to Mott Redfearn, 3 January 1940; all in "North Carolina-M–Z" folder, box 690, Correspondence (1939–40), RG 33.

63. Larkins, *Negro Population of North Carolina*, 20.

64. Dazelle Foster Lowe, "1938 Annual Narrative Report," pp. 6, 11, reel 84, *Annual Reports*; Dazelle Foster Lowe, "1937 Annual Narrative Report," pp. 11–12, reel 77, *Annual Reports*; Dazelle Foster Lowe, "1940 Annual Narrative Report," p. 17, reel 104, *Annual Reports*. This material became the boilerplate for Lowe's annual reports. For example, see Lowe, "1941 Annual Narrative Report," pp. 13–14, reel 113, *Annual Reports*.

65. DeLaine interview, Special Collections, J. Y. Joyner Library, East Carolina University; Wilhelmina R. Laws, "1940 Annual Report," pp. 69–70, reel 103, *Annual Reports*.

66. DeLaine interview, tape 1, side B.

67. Lillian H. Andrews, Home Demonstration Annual Report, Bertie County, 1937, p. 12, reel 78, *Annual Reports*. For similar accounts, see Marietta Meares, Home Demonstration Annual Report, Craven County, 1937, p. 17, reel 79, *Annual Reports*; Wilhelmina R. Laws, "1937 Annual Report," p. 11, reel 77, *Annual Reports*.

68. Dazelle Foster Lowe, "1937 Annual Narrative Report," p. 2, reel 77, *Annual Reports*.

69. Ruth Current, "North Carolina Home Demonstration Leader Annual Report 1940," p. 164, reel 104, *Annual Reports*; Dazelle Foster Lowe, "1940 Annual Narrative Report," p. 7, reel 104, *Annual Reports*.

70. "Negroes Seek to Retain a Home Agent," Goldsboro *News-Argus*, 21 July 1932, p. 1; "Farm-Home Agents Not to Regain Jobs," Goldsboro *News-Argus*, p. 1; and "Negroes Form Agricultural Booster Group," Goldsboro *News-Argus*, 25 July 1932, pp. 1, 3; Dazelle Foster Lowe, "Eighth Annual Report of Extension Work as Conducted by the Negro Home Demonstration Agents of North Carolina, December 1, 1931–December 1, 1932," p. 13.

71. Dazelle Foster Lowe, "Home Demonstration Leader (Negro) Annual Report 1936," p. 5, reel 70, *Annual Reports*; Wilhelmina R. Laws, "1937 Annual Report," p. 45, reel 77, Annual Reports. For a similar description of black home agents' community roles in South Carolina, see Harris, "'Fairy Godmothers' and 'Magicians.'"

Chapter Six

1. "Most Everybody Had Something Made of Feed Sacks," 213.

2. This chapter builds upon research conducted with Sunae Park of the Division of Textiles, National Museum of American History (NMAH), Washington, D.C., between 1989 and 1991. Our collaboration resulted in an exhibit at NMAH, "From Feed Bags to Fashion," August–December 1991, and an article of the same title. For another serious consideration of reusing sacks, see Strasser, *Waste and Want*, 211–15.

3. Cook, *Identification and Value Guide to Textile Bags*, 1–12; Nickols, "The Use of Cotton Sacks in Quiltmaking."

4. Agee and Evans, *Let Us Now Praise Famous Men*; Marion Post Wolcott, "Vegetable pickers, migrants, waiting to be paid after work," February 1939, Homestead, Fla. (vicinity), LC-USF-33-30491-M4, Farm Security Administration/Office of War Information Collection, Division of Prints and Photographs, Library of Congress, Washington, D.C.; Newman, "Making Do."

5. Anonymous poet quoted in Cook, *Identification and Value Guide to Textile Bags*, 2.

6. (Miss) Gay B. Shepperson to Miss Lucy Poindexter, 30 April 1934, Federal Emergency Relief Association, State Files 1933–36, Georgia, RG 69, National Archives, Washington, D.C.; Flynt, *Poor but Proud*, 303; Jane S. McKimmon, "Annual Report, 1931, Home Demonstration Division," pp. 74, 81, "1931" folder, box 5, Jane S. McKimmon Papers, North Carolina Division of Archives and History, Raleigh, N.C. Thanks to Sarah Wilkerson-Freeman for the information from Georgia.

7. Edythe H. Jones, letter to author, 17 January 1990, author's collection; Flynt, *Poor but Proud*, 218–19.

8. Fleming interview, 50, "An Oral History of Southern Agriculture," National Museum of American History, Smithsonian Institution, Washington, D.C. (hereafter cited as OHSA).

9. "Most Everybody Had Something Made of Feed Sacks," 209.

10. Edythe H. Jones, letter to author, 17 January 1990.

11. Lillie H. Hester, Home Demonstration Annual Report, Bladen County, 1937, p. 15a, *Cooperative Extension Service Annual Reports*, reel 78, Microforms Collection, North Carolina State University Archives, D. H. Hill Library, Raleigh, N.C.

12. Roberts interview, 25–26, OHSA.

13. Millers' National Federation, *Milling Around in Washington*, 29 December 1941, pp. 1–3; Cook, *Identification and Value Guide to Textile Bags*, 7–8.

14. Cook, *Identification and Value Guide to Textile Bags*, 9–10.

15. Fleming interview, 49–50, OHSA.

16. "Most Everybody Had Something Made of Feed Sacks," 208.

17. Byers interview, 15–16, OHSA.

18. *Feedstuffs*, July 14, 1945, p. 35.

19. Perkins, "Fashions in Feed Bags," p. 32.

20. Allen, "Feed Bags de Luxe," p. 111.

21. Edythe H. Jones, letter to author, 17 January 1990.

22. *Feedstuffs*, 1 March 1947, p. 43.

23. Ibid., 24 July 1948, p. 55.

24. Ibid., 10 January 1959, p. 65.

25. *Flour and Feed*, January 1948, p. 37.

26. *Feedstuffs*, 15 July 1944, p. 32.

27. Ibid., 14 July 1945, p. 35.

28. Ibid., 14 September 1948, p. 25.

29. *Flour and Feed*, January 1948, p. 37.

30. *Feedstuffs*, 8 February 1947, p. 35.

31. Ibid., 21 June 1947, p. 17.

32. Ibid., 10 July 1948, p. 13.

33. Ibid., 1 February 1947, p. 67.

34. Ibid., 3 July 1948, p. 8.

35. Ibid., 31 July 1948, p. 57.

36. Ibid., 4 December 1948, p. 52.

37. Ibid., 26 March 1949, p. 48.

38. *Smart Sewing with Bags*, 12.

39. *Sew Easy with Cotton Bags*, 12.

40. Carson, "It's in the Bag," pp. 18–19.

41. Cook, *Identification and Value Guide to Textile Bags*, 149–50.

42. Anna Lue Cook supplied a copy of the advertisement for the 1959 Cotton Bag Sewing Contest.

43. *Feedstuffs*, 14 July 1945, p. 35, and 31 July 1948, p. 40.

44. Wheless, "500 Creative Farm Women Display Glamour," p. 8.

45. Fleming interview, 51, OHSA.

BIBLIOGRAPHY

Manuscript Collections

Auburn, Alabama
University Archives, Ralph Brown Draughon Library, Auburn University
  Mary Wigley Harper Collection

Boone, North Carolina
Appalachian Cultural Museum, Appalachian State University
  John Ward Store Collection

Chapel Hill, North Carolina
North Carolina Collection, Louis Round Wilson Library, University of North Carolina
    at Chapel Hill
  Wiley B. Sanders, "Race Attitudes of County Officials"
Southern Historical Collection, Louis Round Wilson Library, University of North
    Carolina at Chapel Hill
  Federal Writers' Project Papers
  Jane Simpson McKimmon Collection

College Park, Maryland
National Archives and Records Administration
  Agricultural Adjustment Administration, Record Group 145
  Extension Service, Record Group 33
  U.S. Department of Agriculture, Record Group 16-G

Durham, North Carolina
Rare Book, Manuscript, and Special Collections, William R. Perkins Library, Duke
    University
    Mack Ivey Cline Memoirs
    Anne Firor Scott Collection

Greenville, North Carolina
Edythe H. Jones Papers, in possession of author

North Wilkesboro, North Carolina
Wilkes Community College Library
    Vertical files

Raleigh, North Carolina
North Carolina Division of Archives and History
    Department of Public Instruction, Division of Negro Education
    Governor O. Max Gardner Papers
    Jane Simpson McKimmon Papers
    Anne Pauline Smith and Frank O. Alford Papers
University Archives, D. H. Hill Library, North Carolina State University
    Agriculture School, Agricultural Extension Service
    Microforms Collection, *Cooperative Extension Service Annual Reports*

Smithfield, North Carolina
Johnston County Heritage Center, Smithfield Public Library
    Mrs. Mamie Sue Jones Evans Papers

Tuskegee, Alabama
Hollis Burke Frissell Library, Tuskegee University
    Robert Russa Moton Papers
    Margaret Murray Washington Papers

Washington, D.C.
Archives Center, National Museum of American History
    Warshaw Collection
Division of Prints and Photographs, Library of Congress
    Farm Security Administration/Office of War Information Collection
Manuscript Division, Library of Congress
    Moton Family Papers

Collections on Microform
Hampton University Newspaper Clippings File
Meier, August, and Elliot Rudwick, eds. *The Claude A. Barnett Papers* (Frederick, Md.:
University Publications of America, Inc.)

Interviews

Durham, North Carolina
Rare Book, Manuscript, and Special Collections, Perkins Library, Duke University
Nelson, Margaret Christine. Interview by Mary Hebert, Summerton, S.C., 5 July
1995, "Behind the Veil: Documenting African-American Life in the Jim Crow
South."

Greenville, North Carolina
Special Collections, J. Y. Joyner Library, East Carolina University
DeLaine, Lucy T. Interview by Lu Ann Jones, 4 March 1999.

North Wilkesboro, North Carolina
Wilkes Community College Library
Lovette, Mrs. Charles O. Videotaped interview by unidentified interviewer, no place,
no date.

Washington, D.C.
National Museum of American History, Smithsonian Institution. "An Oral History of
Southern Agriculture." Interviews by Lu Ann Jones.
Anderson, Walter and Adra. Corryton, Tenn., 3 May 1987.
Benton, Aubrey and Ina Belle. Commerce, Ga., 28 April 1987.
Blackstock, Tom and Velva. Talmo, Ga., 22 April 1987.
Booth, N. J. and Jarutha. Bassfield, Miss., 27 October 1987.
Byers, Ruby. Gainesville, Ga., 23 April 1987.
Cunningham, Tom. Darlington, S.C., 8 January 1987.
Davis, Fredda. Laurel Springs, N.C., 26 May 1987.
Dillard, John T. and Nioma. Tifton, Ga., 22 January 1987.
Felknor, Jessie Franklin. White Pine, Tenn., 2 May 1987.
Fleming, Arthur B. Gainesville, Ga., 27 April 1987.
Foster, Jim and Virgie. Millers Creek, N.C., 19 May 1987.
Gosney, Jessie and Kenneth. Carlisle, Ark., 1 October 1987.
Griffin, A. C. and Grace. Edenton, N.C., 8 and 10 December 1986.
Kilby, T. H. North Wilkesboro, N.C., 19 May 1987.
Langley, Nellie Stancil. Stantonsburg, N.C., 5 December 1986.

Minchew, Edna. Wray, Ga., 21 January 1987.
Mohamed, Ethel Wright. Belzoni, Miss., 23 October 1987.
Murray, Lurline Stokes. Florence, S.C., 6 January, 9 January, 11 January, and 16 January 1987.
Patterson, Vanona. Hiddenite, N.C., 13 May 1987.
Pender, Bessie Smith. Tarboro, N.C., 30 December 1986.
Player, C. B., Jr. Bishopville, S.C., 12 January 1987.
Purvis, Clyde. Tifton, Ga., 26 January 1987.
Redmond, Virgie St. John. Statesville, N.C., 14 May and 29 December 1987.
Roberts, James and Gerti. Roaring River, N.C., 21 May 1987.
Rountree, G. Emory. Sunbury, N.C., 16 December 1986.
Sarten, Della. Sevierville, Tenn., 1 May 1987.
Spivey, Wayland. Edenton, N.C., 9 December 1986.
Thompson, Mioma. Tifton, Ga., 24 January 1987.
Turner, Mary Wiggins. Gatesville, N.C., 16 December 1986.
Welborn, S. L. Jefferson, Ga., 27 April 1987.
Winski, Dent and Arnalee. Tifton, Ga., 23 January 1987.
Young, Walter and Hattie. Bastile, La., 10 May 1988.

Author's Personal Collection

Unless otherwise noted, interviews are by the author, and tape recordings or notes of interviews are in author's possession.

Edmonds, Hal. Mars Hill, N.C., 4 August 1994.
Hubbard, Charlotte Moton. Chevy Chase, Md., 23 October 1993 and 18 March 1994.
Patterson, Catherine Moton. Exeter, N.H., 11 June 1994.
Thomas, Gladys Hackney. Pittsboro, N.C., 17 November 1994.
Vanlandingham, Eugenia Patterson. Tarboro, N.C., 20 August 1982.
Vanlandingham, Eugenia Patterson. Interview by D. W. Colvard, Tarboro, N.C., 10 January 1980. Transcript in author's possession.

Government Documents

Anderson, W. A. "Farm Family Living among White Owner and Tenant Operators in Wake County, 1926." North Carolina Agricultural Experiment Station, *Bulletin* No. 269. September 1929.
Bulletin of the North Carolina Department of Agriculture, February 1900.
Bulletin of the North Carolina Department of Agriculture, January 1914.
Evans, J. A. "Extension Work among Negroes Conducted by Negro Agents, 1923."

U.S. Department of Agriculture, Department *Circular* 355. Washington, D.C.: Government Printing Office, September 1925.

Flohr, Lewis B. "Marketing Farm Produce by Parcel Post." *Farmers' Bulletin*, No. 1551, U.S. Department of Agriculture. Washington, D.C.: Government Printing Office, May 1933.

Lewis, V. W., F. W. Risher, and L. C. Salter. "Carlot Marketing of Poultry in North Carolina." *Bulletin of the North Carolina Department of Agriculture*, October 1927.

Slocum, Rob R. "Marketing Eggs." *Farmers' Bulletin*, No. 1378, U.S. Department of Agriculture. Washington, D.C.: Government Printing Office, June 1928.

Taylor, Carl C., and C. C. Zimmerman. "Economic and Social Conditions of North Carolina Farmers." [Raleigh: State Board of Agriculture, 1923].

U.S. Bureau of the Census. *Twelfth Census of the United States, Taken in the Year 1900*. Vol. 5: *Agriculture, Part 1, Farms, Livestock, and Animal Products*. Washington, D.C.: Government Printing Office, 1902.

U.S. Department of Agriculture, Office of the Secretary. Report No. 103. *Social and Labor Needs of Farm Women*. Washington, D.C.: Government Printing Office, 1915.

————. Report No. 104. *Domestic Needs of Farm Women*. Washington, D.C.: Government Printing Office, 1915.

————. Report No. 106. *Economic Needs of Farm Women*. Washington, D.C.: Government Printing Office, 1915.

U.S. Department of Commerce, Bureau of the Census. *Thirteenth Census of the United States Taken in the Year 1910*. Vol. 5: *Agriculture, 1909 and 1910, General Report and Analysis*. Washington, D.C.: Government Printing Office, 1913.

————. *Fourteenth Census of the United States Taken in the Year 1920*. Vol. 4: *Population, 1920, Occupations*. Washington, D.C.: Government Printing Office, 1923.

————. *Fifteenth Census of the United States: 1930, Agriculture*. Vol. 4: *General Report, Statistics by Subjects*. Washington, D.C.: Government Printing Office, 1932.

————. *Fifteenth Census of the United States: 1930, Agriculture*. Vol. 2: *Part 2: The Southern States*. Washington, D.C.: Government Printing Office, 1932.

————. *Fifteenth Census of the United States: 1930, Population*. Vol. 4: *Occupations by State*. Washington, D.C.: Government Printing Office, 1933.

————. *Sixteenth Census of the United States: 1940, Agriculture*. Vol. 3: *General Report, Statistics by Subjects*. Washington, D.C.: Government Printing Office, 1943.

————. *Sixteenth Census of the United States: 1940, Population*. Vol. 2: *Characteristics of the Population*. Washington, D.C.: Government Printing Office, 1943.

U.S. Department of the Interior, Census Office. *Compendium of the Eleventh Census: 1890*. Part 3. Washington, D.C.: Government Printing Office, 1897.

Periodicals

Broiler Growing
Feedstuffs
Flour and Feed
News-Argus (Goldsboro, N.C.)
Progressive Farmer

Pamphlets

Cobb, John Howell. "Address on Poultry by Hon. Howell Cobb, Delivered before the Georgia State Agricultural Convention, Held at Brunswick, Ga., February 12th, 1889." Rare Book Collection, Southern Pamphlet 6096, Louis Round Wilson Library, University of North Carolina at Chapel Hill.
Sew Easy with Cotton Bags. Memphis, Tenn.: National Cotton Council, 1950.
Smart Sewing with Bags. Memphis, Tenn.: National Cotton Council, 1949.

Bulletins

Bulletin of the Agricultural & Technical College of North Carolina, Twenty-Ninth Annual Catalogue, 1933–1934. Vol. 25. Greensboro, May 1934.
Bulletin of the Agricultural & Technical College of North Carolina, Thirty-Second Annual Catalogue, 1936–1937. Vol. 28. Greensboro, May 1937.

Books and Articles

Abrams, Douglas Carl. "Irony of Reform: North Carolina Blacks and the New Deal." *North Carolina Historical Review* 66 (April 1989): 149–78.
Absher, Mrs. W. O., ed. *The Heritage of Wilkes County, 1982.* Winston-Salem, N.C.: Wilkes Genealogical Society and Hunter Publishing Co., 1982.
Adams, Jane, ed. *All Anybody Every Wanted of Me Was to Work: The Memories of Edith Bradley Rendleman.* Carbondale and Edwardsville: Southern Illinois University Press, 1996.
Agee, James, and Walker Evans. *Let Us Now Praise Famous Men.* Boston: Houghton Mifflin, 1941.
Allen, Gertrude. "Feed Bags de Luxe." *Reader's Digest* (March 1942): 111.
Allen, Ruth Alice. *The Labor of Women in the Production of Cotton.* University of Texas *Bulletin,* No. 3134, 1931. Reprint, New York: Arno Press, 1975.
Ambrose, Linda M., and Margaret Kechnie. "Social Control or Social Feminism: Two Views of the Ontario Women's Institute." *Agricultural History* 73 (Spring 1999): 222–37.

Anderson, James D. *The Education of Blacks in the South, 1860–1935.* Chapel Hill: University of North Carolina Press, 1981.

Atherton, Lewis E. "Itinerant Merchandising in the Antebellum South." *Bulletin of the Business Historical Society* 19 (April 1945): 35–59.

Ayers, Edward L. "Narrating the New South." *Journal of Southern History* 61 (August 1995): 555–66.

———. *The Promise of the New South: Life after Reconstruction.* New York and Oxford: Oxford University Press, 1992.

Babbitt, Kathleen R. "The Productive Farm Woman and the Extension Home Economist in New York State, 1920–1940." In *American Rural and Farm Women in Historical Perspective*, edited by Joan M. Jensen and Nancy Grey Osterud, 83–101. Washington, D.C.: Agricultural History Society, 1994.

Bailey, Joseph Cannon. *Seaman A. Knapp: Schoolmaster of American Agriculture.* Columbia University Studies in the History of American Agriculture, no. 10. New York: Columbia University Press, 1945.

Baker, Gladys. *The County Agent.* Chicago: University of Chicago Press, 1939.

Barrett, Mrs. John W. (Mary X.), ed. *The History of Stephenson County, 1970.* Freeport, Ill.: County of Stephenson, 1972.

Barron, Hal S. *Mixed Harvest: The Second Great Transformation in the Rural North, 1870–1930.* Chapel Hill: University of North Carolina Press, 1997.

Benjamin, Earl W., and Howard C. Pierce. *Marketing Poultry Products.* 3d ed. New York: John Wiley and Sons, 1937.

Benson, Susan Porter. "Living on the Margin: Working-Class Marriages and Family Survival Strategies in the United States, 1919–1941." In *The Sex of Things: Gender and Consumption in Historical Perspective*, edited by Victoria de Grazia, with Ellen Furlough, 212–43. Berkeley: University of California Press, 1996.

Bohannon, Paul, and George Dalton, eds. *Markets in Africa.* [Evanston, Ill.]: Northwestern University Press, 1962.

Bourke, Joanna. "Dairywomen and Affectionate Wives: Women in the Irish Dairy Industry, 1890–1914." *Agricultural History Review* 38, pt. 2 (1990): 149–64.

———. "Women and Poultry in Ireland, 1891–1914." *Irish Historical Studies* 25, no. 99 (May 1987): 293–310.

Bowers, William L. *The Country Life Movement in America, 1900–1920.* Port Washington, N.Y.: Kennikat Press, 1974.

Boydston, Jeanne. *Home and Work: Housework, Wages, and the Ideology of Labor in the Early Republic.* New York and Oxford: Oxford University Press, 1990.

Breen, William J. "Southern Women in the War: The North Carolina Woman's Committee, 1917–1919." *North Carolina Historical Review* 55 (July 1978): 251–81.

Brown, Hugh Victor. *E-Qual-ity Education in North Carolina among Negroes.* Raleigh: Irving Swain Press, 1964.

———. *A History of the Education of Negroes in North Carolina.* Raleigh: Irving Swain Press, 1961.

Bunch, Mamie. "A Course for Home Demonstration Agents: The Illinois Plan." *Journal of Home Economics* 11 (October 1919): 430–35.

Business Opportunities for the Home Economist: New Jobs in Consumer Service. A study made by the Institute of Women's Professional Relations under the direction of Chase Going Woodhouse, director, with the assistance of Anne Evans and the workers on Connecticut WPA Project 2085. McGraw-Hill Euthenics Series. New York and London: McGraw-Hill Book Co., 1938.

Calvin, Henrietta. "Extension Work." *Journal of Home Economics* 9 (December 1917): 565–66.

Campbell, Thomas Monroe. *The Movable School Goes to the Negro Farmer.* Tuskegee Institute, Ala.: Tuskegee Institute Press, 1936. Reprint, New York: Arno Press, 1969.

Carson, Ruth. "It's in the Bag." *Colliers* (4 May 1946): 18–19.

Chandler, Alfred D., Jr. *The Visible Hand: The Managerial Revolution in American Business.* Cambridge, Mass.: Belknap Press of Harvard University Press, 1977.

Clark, Thomas D. *Pills, Petticoats, and Plows: The Southern Country Store.* Norman: University of Oklahoma Press, 1944.

Clinton, Catherine, ed. *Half-Sisters of History: Southern Women and the American Past.* Durham, N.C.: Duke University Press, 1994.

Coclanis, Peter A., and David L. Carlton, eds. *Confronting Southern Poverty in the Great Depression: The Report on Economic Conditions in the South, with Related Documents.* Boston: Bedford Books of St. Martin's Press, 1996.

Cohen, Lizabeth. "The Class Experience of Mass Consumption: Workers as Consumers in Interwar America." In *The Power of Culture: Critical Essays in American History*, edited by Richard Wightman Fox and T. J. Jackson Lears, 135–60. Chicago: University of Chicago Press, 1993.

———. "Encountering Mass Culture at the Grassroots: The Experience of Chicago Workers in the 1920s." *American Quarterly* 41 (March 1989): 6–33.

Coleman, Caroline S. *Five Petticoats on Sunday.* Greenville, S.C.: Hiott, 1962.

Coltman, Robert. "A '90s Murder Mystery: 'The Peddler and His Wife.'" *Old Time Music* 30 (Autumn 1978): 13–15.

Conrad, David Eugene. *The Forgotten Farmers: The Story of Sharecroppers in the New Deal.* Urbana: University of Illinois Press, 1965.

Conway, Jill Ker. *When Memory Speaks: Reflections on Autobiography.* New York: Alfred A. Knopf, 1998.

Cook, Anna Lue. *Identification and Value Guide to Textile Bags.* Florence, Ala.: Books Americana, 1990.

Cookingham, Mary E. "Combining Marriage, Motherhood, and Jobs before World

War II: Women College Graduates, Classes of 1905–1935." *Journal of Family History* 9 (Summer 1984): 178–95.

Cooley, Rosa B. *Homes of the Freed.* With an introduction by J. H. Dillard. New York: New Republic, 1926.

Cott, Nancy F. *The Grounding of Modern Feminism.* New Haven, Conn.: Yale University Press, 1987.

Cotten, Sallie Southall. *History of the North Carolina Federation of Women's Clubs, 1901–1925.* Raleigh: Edwards & Broughton Printing Co., 1925.

Cotton, Barbara R. *The Lamplighters: Black Farm and Home Demonstration Agents in Florida, 1915–1965.* Tallahassee, Fla.: U.S. Department of Agriculture and Florida Agricultural and Mechanical University, 1982.

Crews, Harry. *A Childhood: The Biography of a Place.* New York: Harper and Row, 1978.

Crosby, Earl W. "The Struggle for Existence: The Institutionalization of the Black County Agent System." *Agricultural History* 60 (Spring 1986): 123–36.

Crow, Jeffrey J., Paul D. Escott, and Flora J. Hatley. *A History of African Americans in North Carolina.* Raleigh: Division of Archives and History, North Carolina Department of Cultural Resources, 1992.

Daniel, Pete. *Breaking the Land: The Transformation of Cotton, Tobacco, and Rice Cultures since 1880.* Urbana: University of Illinois Press, 1985.

Day, Kay Young. "Kinship in a Changing Economy: A View from the Sea Islands." In *Holding on to the Land and the Lord: Essays on Kinship, Ritual, Land Tenure, and Social Policy,* edited by Robert L. Hall and Carol B. Stack, 11–24. Athens: University of Georgia Press, 1982.

Edwards, Laura F. *Scarlett Doesn't Live Here Anymore: Southern Women in the Era of the Civil War.* Urbana: University of Illinois Press, 2000.

Ellenberg, George B. "'May the Club Work Go on Forever': Home Demonstration and Rural Progressivism in 1920s Ballard County." *The Register of the Kentucky Historical Society* 96 (Spring 1998): 137–66.

Evans, Sara M. *Born for Liberty: A History of Women in America.* New York: The Free Press, 1989.

Fairclough, Adam. "'Being in the Field of Education and Also Being a Negro . . . Seems . . . Tragic': Black Teachers in the Jim Crow South." *Journal of American History* 87 (June 2000): 65–91.

Faragher, John Mack. "History from the Inside-Out: Writing the History of Women in Rural America." *American Quarterly* 33 (Winter 1981): 537–57.

———. *Sugar Creek: Life on the Illinois Prairie.* New Haven, Conn.: Yale University Press, 1986.

Faulkner, William. *Light in August.* 1932. Reprint, New York: Modern Library, 1968.

Fausto-Sterling, Anne. Review of *Nature's Body: Gender in the Making of Modern Science,* by

Londa Schiebinger; *Gender on Ice: American Ideologies of Polar Expeditions*, by Lisa Bloom; and *A Question of Identity: Women, Science, and Literature*, by Marina Benjamin. *Signs* 21 (Autumn 1995): 172–75.

Filene, Peter G. *Him/Her/Self: Gender Identities in Modern America*. 3d ed. With a foreword by Elaine Tyler May. Baltimore, Md.: Johns Hopkins University Press, 1998.

Fink, Deborah. *Agrarian Women: Wives and Mothers in Rural Nebraska, 1880–1940*. Chapel Hill: University of North Carolina Press, 1992.

———. *Open Country, Iowa: Rural Women, Tradition and Change*. Albany: State University of New York Press, 1986.

———. "Sidelines and Moral Capital: Women on Nebraska Farms in the 1930s." In *Women and Farming: Changing Roles, Changing Structures*, edited by Wava G. Haney and Jane B. Knowles, 55–72. Boulder, Colo.: Westview Press, 1988.

Fite, Gilbert C. *Cotton Fields No More: Southern Agriculture, 1865–1980*. Lexington: University Press of Kentucky, 1984.

Flora, Cornelia Butler, and John Stitz. "Female Subsistence Production and Commercial Farm Survival among Settlement Kansas Wheat Farmers." *Human Organization: Journal of the Society for Applied Anthropology* 47 (Spring 1988): 64–69.

Flynt, J. Wayne. *Poor but Proud: Alabama's Poor Whites*. Tuscaloosa: University of Alabama Press, 1989.

Frisch, Michael. *A Shared Authority: Essays on the Craft and Meaning of Oral and Public History*. Albany: State University of New York Press, 1990.

Frysinger, Grace E. "The Home Economics Extension of the Future." *Journal of Home Economics* 15 (October 1923): 543–45.

Gavins, Raymond. "Fear, Hope, and Struggle: Recasting Black North Carolina in the Age of Jim Crow." In *Democracy Betrayed: The Wilmington Race Riot of 1898 and Its Legacy*, edited by David S. Cecelski and Timothy B. Tyson, 185–206. Chapel Hill: University of North Carolina Press, 1998.

Golden, Harry Lewis. *Forgotten Pioneer*. Cleveland and New York: The World Publishing Co., 1963.

Gordon, Linda. "Black and White Visions of Welfare: Women's Welfare Activism, 1890–1945." *Journal of American History* 78 (September 1991): 559–90.

Grobman, Neil R. "Peddler Black of Owen County: 'Not a Stranger to Us.'" *Kentucky Folklore Record* 24 (January–March 1978): 2–5.

Hagood, Margaret Jarman. *Mothers of the South: Portraiture of the White Tenant Farm Woman*. Chapel Hill: University of North Carolina Press, 1939. Reprint, New York: W. W. Norton & Co., 1977.

Hahamovitch, Cindy. *The Fruits of Their Labor: Atlantic Coast Farmworkers and the Making of Migrant Poverty, 1870–1945*. Chapel Hill: University of North Carolina Press, 1997.

Hahn, Steven. *The Roots of Southern Populism: The Transformation of the Georgia Upcountry, 1850–1890*. New York: Oxford University Press, 1982.

Hale, Grace Elizabeth. *Making Whiteness: The Culture of Segregation in the South, 1890–1940.* New York: Pantheon Books, 1998.

Hall, Jacquelyn Dowd. "Partial Truths: Writing Southern Women's History." *Signs* 14 (Summer 1989): 902–11.

Hall, Jacquelyn Dowd, and Anne Firor Scott. "Women in the South." In *Interpreting Southern History: Historiographical Essays in Honor of Sanford W. Higginbotham,* edited by John B. Boles and Evelyn T. Nolen, 454–509. Baton Rouge: Louisiana State University Press, 1987.

Hamby, Zetta Barker. *Memoirs of Grassy Creek: Growing Up in the Mountains on the Virginia–North Carolina Line.* Jefferson, N.C.: McFarland & Co., 1998.

Harris, Bernice Kelly. *Folk Plays of Eastern Carolina.* Edited with an introduction by Frederick H. Koch. Chapel Hill: University of North Carolina Press, 1940.

———. *Purslane.* Chapel Hill: University of North Carolina Press, 1939.

———. *Southern Savory.* Chapel Hill: University of North Carolina Press, 1964.

———. *Sweet Beulah Land.* Garden City, N.Y.: Doubleday, Doran and Co., 1943.

Harris, Carmen. "Grace under Pressure: The Black Home Extension Service in South Carolina, 1919–1966." In *Rethinking Home Economics: Women and the History of a Profession,* edited by Sarah Stage and Virginia B. Vincenti, 203–28. Ithaca, N.Y.: Cornell University Press, 1997.

Higginbotham, Evelyn Brooks. *Righteous Discontent: The Women's Movement in the Black Baptist Church, 1880–1920.* Cambridge, Mass.: Harvard University Press, 1993.

Hilton, Kathleen C. "'Both in the Field, Each with a Plow': Race and Gender in USDA Policy, 1907–1929." In *Hidden Histories of Women in the New South,* edited by Virginia Bernhard, Betty Brandon, Elizabeth Fox-Genovese, Theda Perdue, and Elizabeth Hayes Turner, 114–33. Columbia: University of Missouri Press, 1994.

Hine, Darlene Clark. *Hine Sight: Black Women and the Re-Construction of American History.* Brooklyn, N.Y.: Carlson Publishing, 1994.

Hobbs, S. H., Jr. *Know Your Own State—North Carolina.* Chapel Hill: University of North Carolina Press, 1924.

———. *North Carolina, Economic and Social.* Chapel Hill: University of North Carolina Press, 1930.

Hoffschwelle, Mary S. *Rebuilding the Rural Southern Community: Reformers, Schools, and Homes in Tennessee, 1900–1930.* Knoxville: University of Tennessee Press, 1998.

———. "The Science of Domesticity: Home Economics at George Peabody College for Teachers, 1914–1939." *Journal of Southern History* 57 (November 1991): 659–80.

Hollingsworth, Laura, and Vappu Tyyska. "The Hidden Producers: Women's Household Production during the Great Depression." *Critical Sociology* 15 (Fall 1988): 3–27.

Holt, Marilyn Irvin. *Linoleum, Better Babies, and the Modern Farm Woman, 1890–1930.* Albuquerque: University of New Mexico Press, 1995.

Holt, Sharon Ann. *Making Freedom Pay: North Carolina Freedpeople Working for Themselves, 1865–1900*. Athens: University of Georgia Press, 2000.

Jaffee, David. "Peddlers of Progress and the Transformation of the Rural North, 1760–1860." *Journal of American History* 78 (September 1991): 511–35.

Jellison, Katherine. *Entitled to Power: Farm Women and Technology, 1913–1963*. Chapel Hill: University of North Carolina Press, 1993.

Jensen, Joan M. "Butter Making and Economic Development in Mid-Atlantic America, 1750–1850." In *Promise to the Land: Essays on Rural Women*, 170–85. Albuquerque: University of New Mexico Press, 1991.

———. "Cloth, Butter, and Boarders: Women's Household Production for Market." In *Promise to the Land: Essays on Rural Women*, 186–205. Albuquerque: University of New Mexico Press, 1991.

———. *Loosening the Bonds: Mid-Atlantic Farm Women, 1750–1850*. New Haven, Conn.: Yale University Press, 1987.

———. *Promise to the Land: Essays on Rural Women*. Albuquerque: University of New Mexico Press, 1991.

———. "Recovering Her Story: Learning the History of Farm Women." In *Promise to the Land: Essays on Rural Women*, 73–82. Albuquerque: University of New Mexico Press, 1991.

———. *With These Hands: Women Working on the Land*. Old Westbury, N.Y.: Feminist Press, 1981.

Jeter, Frank, Jr. "'Sitting Up' with the Fire Often Inspired Romance." *Carolina Country* (November 1988): 20.

Johnson, F. Roy. *Tales of Country Folks Down Carolina Way*. Murfreesboro, [N.C.]: Johnson Publishing Co., 1978.

Jones, Allen W. "The South's First Black Farm Agents." *Agricultural History* 50 (October 1976): 636–44.

Jones, Bessie. *For the Ancestors: Autobiographical Memories*. Collected and edited by John Stewart. Urbana: University of Illinois Press, 1983.

Jones, Jacqueline. *The Dispossessed: America's Underclasses from the Civil War to the Present*. New York: Basic Books, 1992.

———. *Labor of Love, Labor of Sorrow: Black Women, Work, and the Family, from Slavery to the Present*. New York: Basic Books, 1985.

Jones, Lance G. E. *The Jeanes Teacher in the United States, 1908–1933*. Chapel Hill: University of North Carolina Press, 1937.

Jones, Lu Ann. "'Mama Learned Us to Work': An Oral History of Virgie St. John Redmond." *Oral History Review* 17 (Fall 1989): 63–90.

———. "Voices of Southern Agricultural History." In *International Annual of Oral History, 1990: Subjectivity and Multiculturalism in Oral History*, edited by Ronald J. Grele, 134–44. New York and Westport, Conn.: Greenwood Press, 1992.

Jones, Lu Ann, and Nancy Grey Osterud. "Breaking New Ground: Oral History and Agricultural History." *Journal of American History* 76 (September 1989): 551–64.

Jones, Lu Ann, and Sunae Park. "From Feed Bags to Fashion." *Textile History* 24 (Spring 1993): 91–103.

Keith, Jeanette. *Country People in the New South: Tennessee's Upper Cumberland.* Chapel Hill: University of North Carolina Press, 1995.

Kerber, Linda K., and Jane DeHart Mathews. *Women's America: Refocusing the Past.* 4th ed. New York and Oxford: Oxford University Press, 1995.

Kirby, Jack Temple. *Rural Worlds Lost: The American South, 1920–1960.* Baton Rouge: Louisiana State University Press, 1987.

Ladd-Taylor, Molly. *Mother-Work: Women, Child Welfare, and the State, 1890–1930.* Urbana: University of Illinois Press, 1994.

———. "'My Work Came Out of Agony and Grief': Mothers and the Making of the Sheppard-Towner Act." In *Mothers of a New World: Maternalist Politics and the Origins of Welfare States,* edited by Seth Koven and Sonya Michel, 321–42. New York and London: Routledge, 1993.

Larkins, John R. *The Negro Population of North Carolina.* Special Bulletin No. 23. Raleigh: North Carolina State Board of Charities and Public Welfare, 1944.

Lasch-Quinn, Elizabeth. *Black Neighbors: Race and the Limits of Reform in the American Settlement House Movement, 1890–1945.* Chapel Hill: University of North Carolina Press, 1993.

Lears, Jackson. *Fables of Abundance: A Cultural History of Advertising in America.* New York: Basic Books, 1994.

Leloudis, James. *Schooling the New South: Pedagogy, Self, and Society in North Carolina, 1880–1920.* Chapel Hill: University of North Carolina Press, 1995.

Lewis, John, with Michael D'Orso. *Walking with the Wind: A Memoir of the Movement.* New York: Simon and Schuster, 1998.

Link, William A. *The Paradox of Southern Progressivism, 1880–1930.* Chapel Hill: University of North Carolina Press, 1992.

Littlefield, Valinda Rogers. "Annie Welthy Daughtry Holland." In *Black Women in America: An Historical Encyclopedia.* Vol. 1: *A–L,* edited by Darlene Clark Hine, 579–70. Brooklyn, N.Y.: Carlson Publishing Co., 1993.

Lloyd, William A. "Home Economics Extension—Purpose, Progress, and Prospect." *Journal of Home Economics* 18 (January 1926): 13–15.

Lord, Jerry Dennis. "The Growth and Localization of the United States Broiler Chicken Industry." *Southeastern Geographer* 11 (1971): 29–42.

Lyons-Jennings, Cheryl. "A Telling Tirade: What Was the Controversy Surrounding Nineteenth-Century Midwestern Tree Agents Really About?" *Agricultural History* 72 (Fall 1998): 675–707.

McCurry, Stephanie. *Masters of Small Worlds: Yeoman Households, Gender Relations, and the*

Political Culture of the Antebellum South Carolina Low Country. New York: Oxford University Press, 1995.

McKimmon, Jane Simpson. *When We're Green We Grow*. Chapel Hill: University of North Carolina Press, 1945.

McMurry, Sally. *Transforming Rural Life: Dairying Families and Agricultural Change, 1820–1885*. Baltimore, Md.: Johns Hopkins University Press, 1995.

Martin, O. B. *The Demonstration Work*. Boston: Stratford Co., 1921.

———. "Home Demonstration Work." *Journal of Home Economics* 13 (September 1921): 408–12.

May, Martha. "The 'Good Managers': Married Working Class Women and Family Budget Studies, 1895–1915." *Labor History* 25 (Summer 1984): 351–72.

Melosh, Barbara. *"The Physician's Hand": Work, Culture, and Conflict in American Nursing*. Philadelphia: Temple University Press, 1982.

Melvin, Bruce L. "Rural Life." *American Journal of Sociology* 37 (May 1932): 937–41.

Miller, W. H. "Fleming Ready for Competition." *Broiler Growing* (February 1951): 14, 36–37.

Miller's National Federation. *Milling around in Washington* (29 December 1941).

Mohr, Clarence L. *On the Threshold of Freedom: Masters and Slaves in Civil War Georgia*. Athens: University of Georgia Press, 1986.

Morris, Tom. *Poultry Can Crow at NCSU: A History of Poultry at North Carolina State University*. [Raleigh]: Privately published, 1980.

"Most Everybody Had Something Made of Feed Sacks." *Foxfire* 13 (Fall 1979): 206–13.

Muncy, Robyn. *Creating a Female Dominion in American Reform, 1890–1935*. New York: Oxford University Press, 1991.

NASC Interim History Writing Committee. *The Jeanes Story: A Chapter in the History of American Education, 1908–1968*. N.p.: Southern Education Foundation, 1979.

Neth, Mary. *Preserving the Family Farm: Women, Community, and the Foundations of Agribusiness in the Midwest, 1900–1940*. Baltimore, Md.: Johns Hopkins University Press, 1995.

Newbold, N. C. *Five North Carolina Negro Educators*. Chapel Hill: University of North Carolina Press, 1939.

Newman, Joyce Joines. "Making Do." In *North Carolina Quilts*, edited by Ruth Haislip Roberson, 7–35. Chapel Hill: University of North Carolina Press, 1988.

Nickols, Pat L. "The Use of Cotton Sacks in Quiltmaking." In *Uncoverings*, 1988, Vol. 9 of the Research Papers of the American Quilt Study Group, edited by Laurel Horton, 57–71. San Francisco: American Quilt Study Group, 1988.

Nye, Claribel. "The State Home Demonstration Leader and Her Program." *Journal of Home Economics* 18 (August 1926): 431–35.

Odum, Howard W. *Southern Regions of the United States*. Chapel Hill: University of North Carolina Press, 1936.

Osterud, Nancy Grey. *Bonds of Community: The Lives of Farm Women in Nineteenth Century New York*. Ithaca, N.Y.: Cornell University Press, 1991.

Osterud, Nancy Grey, and Lu Ann Jones. "'If I Must Say So Myself': Oral Histories of Rural Women." *Oral History Review* 17 (Fall 1989): 1–23.

Ownby, Ted. *American Dreams in Mississippi: Consumers, Poverty, and Culture, 1830–1998*. Chapel Hill: University of North Carolina Press, 1999.

———. *Subduing Satan: Religion, Recreation, and Manhood in the Rural South*. Chapel Hill: University of North Carolina Press, 1990.

Pendergast, Tom. "Consuming Questions: Scholarship on Consumerism in America to 1940." *American Studies International* 36 (June 1998): 23–43.

Perkins, D. W. "Fashions in Feed Bags." *American Magazine* (March 1948): 32.

Poe, Clarence. *My First 80 Years*. Chapel Hill: University of North Carolina Press, 1963.

Powell, Ola. "Home Demonstration Work in France." *Journal of Home Economics* 16 (April 1924): 171–76.

Pritchett, Katherine A. "The Training of the County Agent." *Journal of Home Economics* 8 (July 1916): 368–70.

Radway, Janice A. *Reading the Romance: Women, Patriarchy, and Popular Literature*. Chapel Hill: University of North Carolina Press, 1984.

Rasmussen, Wayne D. *Taking the University to the People: Seventy-Five Years of Cooperative Extension*. Ames: Iowa State University Press, 1989.

[Rawleigh, W. T.]. *Guide Book to Help Rawleigh Retailers: A Book on Salesmanship*. Freeport, Ill.: W. T. Rawleigh Co., 1921.

Rawleigh, W. T. *Rawleigh Methods: A Guidebook for Rawleigh Customers*. Freeport, Ill.: W. T. Rawleigh Co., 1926.

The Rawleigh Industries. Freeport, Ill.: W. T. Rawleigh Co., 1932.

Rawleigh's 1927 Good Health Guide, Almanac, Cook Book. Freeport, Ill.: W. T. Rawleigh Co., 1926.

Rawleigh's Stock and Poultry Raisers' Guide. Freeport, Ill.: W. T. Rawleigh Co., n.d.

"Recommendations of the Committee on Extension Needs and Maintenance." *Journal of Home Economics* 13 (September 1921): 422–24.

Reeder, R. L. *The People and the Profession: Pioneers and Veterans of the Extension Service Remember How They Did Their Jobs*. N.p.: Epsilon Sigma Phi, 1979.

Reid, Debra. "Rural African Americans and Progressive Reform." *Agricultural History* 74 (Spring 2000): 322–39.

Rieff, Lynne A. "'Go Ahead and Do All You Can': Southern Progressives and Alabama Home Demonstration Clubs, 1914–1940." In *Hidden Histories of Women in the New South*, edited by Virginia Bernhard, Betty Brandon, Elizabeth Fox-Genovese, Theda Perdue, and Elizabeth Hayes Turner, 134–52. Columbia: University of Missouri Press, 1994.

Riegelhaupt, Joyce F. "Saloio Women: An Analysis of Informal and Formal Political and Economic Roles of Portuguese Peasant Women." *Anthropological Quarterly* (3): 109–26.

Robertson, Ben. *Red Hills and Cotton: An Upcountry Memory.* New York: Alfred A. Knopf, 1942.

Rogers, Lou. *Tar Heel Women.* Raleigh, N.C.: Warren Publishing Co., 1949.

Ross, Ellen. *Love and Toil: Motherhood in Outcast London, 1870–1918.* New York and Oxford: Oxford University Press, 1993.

Rossiter, Margaret W. *Women Scientists in America: Struggles and Strategies to 1940.* Baltimore, Md.: Johns Hopkins University Press, 1982.

Rouse, Jacqueline Anne. *Lugenia Burns Hope: Black Southern Reformer.* Athens: University of Georgia Press, 1989.

Sachs, Carolyn E. "The Participation of Women and Girls in Market and Non-Market Activities on Pennsylvania Farms." In *Women and Farming: Changing Structures, Changing Roles,* edited by Wava G. Haney and Jane B. Knowles, 123–34. Boulder, Colo.: Westview Press, 1988.

Salem, Dorothy. "National Association of Colored Women." In *Black Women in America: An Historical Encyclopedia,* edited by Darlene Clark Hine, 842–51. Brooklyn, N.Y.: Carlson Publishing Co., 1993.

Sanderson, Dwight. "Land-Grant Institutions and Rural Social Welfare." *Journal of Home Economics* 27 (March 1935): 143–45.

Sawyer, Gordon. *The Agribusiness Poultry Industry: A History of Its Development.* Jericho, N.Y.: Exposition Press, 1971.

Scharff, Virginia. *Taking the Wheel: Women and the Coming of the Motor Age.* New York: Free Press, 1991.

Schlereth, Thomas J. "Country Stores, County Fairs, and Mail-Order Catalogues: Consumption in Rural America." In *Consuming Visions: Accumulation and Display of Goods in America,* edited by Simon J. Bronner, 339–75. New York: W. W. Norton & Co., 1987. Published for the Henry Francis du Pont Winterthur Museum, Winterthur, Del.

Schmiechen, James, and Kenneth Carls. *The British Market Hall: A Social and Architectural History.* New Haven, Conn.: Yale University Press, 1999.

Schor, Joel. "The Black Presence in the U.S. Cooperative Extension Service Since 1945: An American Quest for Service and Equity." *Agricultural History* 60 (Spring 1986): 137–53.

Schultz, Mark R. "The Dream Realized?: African American Landownership in Central Georgia between Reconstruction and World War Two." *Agricultural History* 72 (Spring 1998): 298–312.

Scott, Anne Firor. *Natural Allies: Women's Associations in American History.* Urbana: University of Illinois Press, 1991.

————. *The Southern Lady: From Pedestal to Politics, 1830–1930.* 1970. Reprint. 25th Anniversary Edition, with a New Afterword. Charlottesville: University of Virginia Press, 1995.

Scott, Roy V. *The Reluctant Farmer: The Rise of Agricultural Extension to 1914.* Urbana: University of Illinois Press, 1970.

Scull, Penrose, with Prescott C. Fuller. *From Peddlers to Merchant Princes: A History of Selling in America.* New York and Chicago: Follett Publishing Co., 1967.

Seals, R. Grant. "The Formation of Agricultural and Rural Development Policy with Emphasis on African-Americans: II. The Hatch-George and Smith-Lever Acts." *Agricultural History* 65 (Spring 1991): 12–34.

Shand, Hope. "Billions of Chickens: The Business of the South." *Southern Exposure* 11 (November/December 1983): 76–82.

Sharpless, Rebecca. *Fertile Ground, Narrow Choices: Women on Cotton Farms of the Texas Blackland Prairie, 1900–1940.* Chapel Hill: University of North Carolina Press, 1999.

Shaw, Stephanie J. *"What a Woman Ought to Be and to Do": Black Professional Women Workers during the Jim Crow Era.* Chicago: University of Chicago Press, 1996.

Shelton, W. F. *The Day the Black Rain Fell.* Louisburg, N.C.: W. F. Shelton, 1984.

"She's 95, but Still Feels Young, Active." *(Taylorsville, N.C.) Times Advantage,* 21 March 1987.

Simmons-Henry, Linda, and Linda Harris Edminston, eds. *Culture Town: Life in Raleigh's African American Communities.* Raleigh, N.C.: Raleigh Historic Districts Commission, 1993.

Simonsen, Thordis, ed. *You May Plow Here: The Narrative of Sara Brooks.* New York: Simon and Schuster, 1986.

Simpson, Mrs. Nancy W., ed. *The Heritage of Wilkes County, North Carolina, 1990,* vol. 2. Charlotte, N.C.: Wilkes Genealogical Society and Delmar Printing and Publishing, 1990.

Sims, Anastatia. *The Power of Femininity in the New South: Women's Organizations and Politics in North Carolina, 1880–1930.* Columbia: University of South Carolina Press, 1997.

Smith, Margaret Supplee, and Emily Herring Wilson. *North Carolina Women: Making History.* Chapel Hill: University of North Carolina Press, 1999.

Smith, Page, and Charles Daniel. *The Chicken Book.* San Francisco: North Point Press, 1982.

Smith, Susan L. *Sick and Tired of Being Sick and Tired: Black Women's Health Activism in America, 1890–1950.* Philadelphia: University of Pennsylvania Press, 1995.

Sommestad, Lena. "Able Dairymaids and Proficient Dairymen: Education and De-Feminization in the Swedish Dairy Industry." *Gender and History* 4 (Spring 1992): 34–48.

Spears, James E. "Rolling Store: A Reminiscence." *Kentucky Folklore Record* 24 (January–March 1978): 20–21.

————. "Where Have All the Peddlers Gone?" *Kentucky Folklore Record* 21 (July–Sept. 1975): 77–81.

Spears, Timothy B. *100 Years on the Road: The Traveling Salesman in American Culture*. New Haven, Conn.: Yale University Press, 1995.

Stage, Sarah. "Introduction: Home Economics, What's in a Name?" In *Rethinking Home Economics: Women and the History of a Profession*, edited by Sarah Stage and Virginia B. Vincenti, 1–13. Ithaca, N.Y.: Cornell University Press, 1997.

Stearns, Peter N. "Stages of Consumerism: Recent Work on the Issues of Periodization." *Journal of Modern History* 69 (March 1997): 102–17.

Strasser, Susan. *Satisfaction Guaranteed: The Making of the American Mass Market*. New York: Pantheon Books, 1989.

————. *Waste and Want: A Social History of Trash*. New York: Metropolitan Books, Henry Holt and Co., 1999.

Taylor, Roy G. *Down a Country Road*. Wilson, N.C.: J-Mark Publishers, 1986.

————. *Sharecroppers: The Way We Really Were*. Wilson, N.C.: J-Mark Publishers, 1984.

Terrill, Tom E., and Jerrold Hirsch. *Such as Us: Southern Voices of the Thirties*. Chapel Hill: University of North Carolina Press, 1978.

Tindall, George B. *The Emergence of the New South, 1913–1945*. A History of the South, vol. 10. Baton Rouge: Louisiana State University Press, 1967.

Trachtenberg, Alan. *The Incorporation of America: Culture and Society in the Gilded Age*. New York: Hill and Wang, 1982.

Ulrich, Laurel Thatcher. *Good Wives: Image and Reality in the Lives of Women in Northern New England, 1650–1750*. New York: Oxford University Press, 1983.

————. *A Midwife's Tale: The Life of Martha Ballard, Based on Her Diary, 1785–1812*. New York: Vintage Books, 1990.

Valenze, Deborah. "The Art of Women and the Business of Men: Women's Work and the Dairy Industry c. 1740–1840." *Past and Present*, no. 130 (February 1991): 142–69.

Vicinus, Martha. "'One Life to Stand Beside Me': Emotional Conflicts in First-Generation College Women in England." *Feminist Studies* 8 (Fall 1982): 603–28.

Walker, Melissa. *All We Knew Was to Farm: Rural Women in the Upcountry South, 1919–1941*. Baltimore, Md.: Johns Hopkins University Press, 2000.

Walkowitz, Daniel J. "The Making of a Feminine Professional Identity: Social Workers in the 1920s." *Journal of American History* 75 (October 1990): 1051–75.

————. *Working with Class: Social Workers and the Politics of Middle-Class Identity*. Chapel Hill: University of North Carolina Press, 1999.

Walser, Richard. *Bernice Kelly Harris: Storyteller of Eastern Carolina*. Chapel Hill: University of North Carolina Library Extension Publications, 1955.

Warren, Maude. "She 'Lifted' Sixty-Six Counties." *Country Gentleman* (29 June 1918).

Watkins, Floyd C., and Charles Hubert Watkins. *Yesterday in the Hills*. Athens: University of Georgia Press, 1973.

Watkins, J. R., Medical Company. *A Merry Christmas and a Happy New Year from the J. R. Watkins Co. and Your Watkins Dealer.* [Winona, Minn.: The J. R. Watkins Medical Co.], 1938.

———. *The Open Door to Success.* Winona, Minn.: The J. R. Watkins Medical Co., [191?].

Weber, Eugen. *Peasants into Frenchmen: The Modernization of Rural France, 1870–1914.* Stanford, Calif.: Stanford University Press, 1976.

Weems, Robert E., Jr. *Desegregating the Dollar: African American Consumerism in the Twentieth Century.* New York: New York University Press, 1998.

Weigley, Emma Seifrit. "It Might Have Been Euthenics: The Lake Placid Conference and the Home Economics Movement." *American Quarterly* 26 (March 1974): 79–96.

Westling, Louise, ed. *He Included Me: The Autobiography of Sarah Rice.* Athens: University of Georgia Press, 1989.

Wheless, Betty. "500 Creative Farm Women Display Glamour, Originality in FCX Fashion Parade Events." *The FCX Patron* (28 August 1948).

White, John H., Jr. "Home to Roost: The Story of Live Poultry Transit by Rail." *Agricultural History* 63 (Summer 1989): 81–94.

Willard, John D. "Recruiting Extension Workers." *Journal of Home Economics* 13 (September 1921): 412–15.

Witz, Anne, and Mike Savage. "The Gender of Organizations." In *Gender and Bureaucracy*, edited by Mike Savage and Anne Witz, 3–61. Oxford and Cambridge, Mass.: Blackwell Publishers, 1992.

Woloch, Nancy. *Women and the American Experience: A Concise History.* New York: McGraw-Hill Companies, 1996.

Wolters, Raymond. *Negroes and the Great Depression: The Problem of Economic Recovery.* Contributions in American History, No. 6. Westport, Conn.: Greenwood Publishing Co., 1970.

Wood, Betty. *Women's Work, Men's Work: The Informal Slave Economies of Lowcountry Georgia.* Athens: University of Georgia Press, 1995.

Woodruff, Nan Elizabeth. "African-American Struggles for Citizenship in the Arkansas and Mississippi Deltas in the Age of Jim Crow." *Radical History Review* 55 (Winter 1993): 33–51.

———. "Pick or Fight: The Emergency Farm Labor Program in the Arkansas and Mississippi Deltas during World War II." *Agricultural History* 64 (Spring 1990): 74–85.

Woodson, Carter Godwin. *The Rural Negro.* Washington, D.C.: Association for the Study of Negro Life and History, 1930.

Woodward, C. Vann. *Origins of the New South, 1877–1913.* A History of the South, vol. 9. Baton Rouge: Louisiana State University Press, 1951.

Wright, Gavin. *Old South, New South: Revolutions in the Southern Economy since the Civil War.* New York: Basic Books, 1986.

Wright, Richardson. *Hawkers and Walkers in Early America: Strolling Peddlers, Preachers, Lawyers, Doctors, Players, and Others, from the Beginning to the Civil War.* Philadelphia: J. B. Lippincott, 1927.

Yow, Valerie Raleigh. *Bernice Kelly Harris: A Good Life Was Writing.* Baton Rouge: Louisiana State University Press, 1999.

Dissertations, Theses, and Unpublished Papers

Arndt, Carroll Brevard. "Locational Considerations in the North Carolina Broiler Industry." M.A. thesis, University of North Carolina at Chapel Hill, 1969.

Benson, Susan Porter. "Consuming Questions: Thoughts on Gender, Class, and Race in the Social History of Consumption." Paper delivered at the annual meeting of the American Historical Association, Cincinnati, 27–30 December 1988.

Dean, Pamela. "Covert Curriculum: Class, Gender, and Student Culture at a New South Woman's College, 1892–1910." Ph.D. diss., University of North Carolina at Chapel Hill, 1994.

Hanson, Susan Atherton. "Home Sweet Home: Industrialization's Impact on Rural Households, 1865–1925." Ph.D. diss., University of Maryland, 1986.

Harris, Carmen. "'Fairy Godmothers' and 'Magicians': Black Home Demonstration Agents in South Carolina, 1919–1967." Paper delivered at the Berkshire Conference on the History of Women, Poughkeepsie, N.Y., June 1993.

Jones, Lu Ann. "Reading Between the Lines: The Annual Report as Literary Text." Paper delivered at the Fifth Conference on Rural/Farm Women in Historical Perspective, Chevy Chase, Md., December 1994.

———. "'The Task That Is Ours': White North Carolina Farm Women and Agrarian Reform, 1886–1914." M.A. thesis, University of North Carolina at Chapel Hill, 1983.

Lewis, V. W., and D. L. James. "A Poultry Products Marketing Survey in the Counties of Macon, Cleveland, Anson, Robeson, and Beaufort, North Carolina." [Raleigh, N.C., 1924], typescript, National Agricultural Library, Beltsville, Md.

McCleary, Ann Elizabeth. "Home Demonstration and Domestic Reform in Rural Virginia, 1900–1940." Ph.D. diss., Brown University, 1996.

———. "The Home Demonstration Club Movement and Domestic Reform on the Farm in Rural Virginia, 1910–1940." Paper delivered at the Berkshire Conference on the History of Women, Poughkeepsie, N.Y., June 1993.

Pettey, Adrienne. "Standing Their Ground: Small Farm Owners in North Carolina's Tobacco Belt, 1925–1975." Ph.D. diss., Columbia University, 2001.

Rainer, Joseph T. "The Honorable Fraternity of Moving Merchants: Yankee Peddlers in the Old South, 1800–1860." Ph.D. diss., College of William and Mary, 1999.

———. "Yankee Peddlers and Antebellum Southern Consumers in the Arena of Exchange." Paper delivered at the Shopping and Trading in Early America Symposium, George Mason University, Fairfax, Va., 14 March 1997.

Reiff, Lynne Anderson. "'Rousing the People of the Land': Home Demonstration Work in the Deep South, 1914–1950." Ph.D. diss., Auburn University, 1995.

Schoen, Johanna. "'A Great Thing for Poor Folks': Birth Control, Sterilization, and Abortion in Public Health and Welfare in the Twentieth Century." Ph.D. diss., University of North Carolina at Chapel Hill, 1995.

Schultz, Mark Roman. "The Unsolid South: An Oral History of Race, Class, and Geography in Hancock County, Georgia, 1910–1950." Ph.D. diss., University of Chicago, 1999.

Watkins, Charles Alan. "The Ward General Store Exhibit." Working paper typescript, n.d. Appalachian Cultural Museum, Appalachian State University, Boone, N.C.

Young, Wade Phillips. "A History of Agricultural Education in North Carolina." Ph.D. diss., University of North Carolina at Chapel Hill, 1934.

North Carolina: land tenure in, 5; agricultural extension service in, 15, 17, 109, 141, 164–65; home demonstration clubs in, 18–19, 22, 109–11, 189 (n. 44); public health in, 20, 154–55; markets in, 63–64; value of farm products in, 74, 84; poultry industry in, 85, 90–91, 103–4; racial attitudes in, 149–50

North Carolina Agricultural and Technical College, 152

North Carolina Council of Defense, 145

North Carolina State Board of Agriculture, 84–85

Nutrition, 21, 143, 144, 155, 167

Open House (Harris), 53–54

Oral history: author's philosophy of, 4

Ownby, Ted, 32

Pair of Quilts (Harris), 39

Patterson, Eugenia, 158–59

Patterson, Vanona, 81–82

Peanuts, 9

Pearson, S. B., 47

Peddlers, 27, 29–32, 34–36, 43, 70. *See also* Itinerant commerce; Itinerant merchants

Pellagra, 20

Pender, Bessie Smith, 5–6

Percy Kent Bag Company, 178–79

Pierce, John B., 141, 161

"Pin money," 76

Player, C. B., Jr., 61, 76

Poe, Clarence, 15–16

Poultry industry, 81–105; distribution network, 58–60; women's role in, 82–83, 85, 93, 102–3, 104–5, 199 (n. 3); commercialization of, 83, 84, 98–99, 101–5; value of, 84–85; and husbandry, 86, 87, 89–90, 92; and women's cooperatives, 95–96; and scientific agriculture, 100; in North Carolina, 103–4; and World War II, 104. *See also* Eggs

Poultry products, 16, 53, 54, 56, 70–72, 83; and bartering, 56–59; and self-provisioning, 90; distribution of, 95, 99, 102; as economic buffer, 98. *See also* Eggs

Poverty, 3–4, 13–14

Progressive Farmer, 15–16, 86–89, 91, 98–101

Public health, 20, 143, 144, 154–55

Purslane (Harris), 40

Purvis, Clyde, 13, 46

Racial etiquette, 16

Racial ideology, 149–50

Racial uplift, 16, 140

Racism. *See* White supremacy

Raleigh Women's Club, 110

Rawleigh Company, W. T., 29, 41–48

Redfearn, Rosalind, 96

Redmond, Mott, 11

Redmond, Virgie St. John, 9–11

Reform, 2, 17, 108–9, 111, 116

Respectability: politics of, 16, 153, 168

Rice, Sarah Webb, 60–61

Roberts, Gerti, 102, 174

Rockefeller, John D., Jr., 17

Rolling stores. *See* Itinerant merchants

Roosevelt, Franklin Delano, 13–14

Roosevelt, Theodore, 14

Rountree, G. Emory, 59–60

Sarten, Della, 39, 76

Sarten, Will, 76

Scott, Anne Firor, 187 (n. 21)

Segregation, 66, 141–42

Sharecroppers, 3, 5–6, 53
Sheppard-Towner Act, 187 (n. 23)
Smith, Anne Pauline, 108; background of, 110–11, 116; home demonstration activities of, 118–20, 134–35; education of, 121–22; as administrator, 123, 135–37; and opinions of farm men, 124, 130–32; and tension between career and courtship, 124–25, 126, 130–34, 138; and Frank O. Alford, 126, 130–34; and political lobbying, 126–28; and travel, 128–29; and service ideal, 129, 138; and Jane McKimmon, 136
Smith-Hughes Act, 116
Smith-Lever Act, 15, 109, 116, 141–42, 187 (n. 23)
Soil Conservation and Domestic Allotment Act, 160
South Carolina, 74
Southern Tenant Farmers' Union, 160
Spivey, Wayland, 48
Stanton, C. H., 146
Stokes, Julia, 7, 49–50, 61, 77, 78
Subsistence strategies, 5–7, 12, 13, 14; on cash-crop farms, 53, 56; and balancing with bartering and selling, 56, 72
Sugar cane, 53

Taylor, Roy, 12–13, 56, 74
Tenancy, 4–5
Tenant farmers, 3, 53, 160
Textile Bag Manufacturers Association, 177, 180, 181
Thomas, Gladys Hackney, 44–45
Thompson, Mioma, 37
Tobacco, 8
Tomato Clubs, 17–18, 109, 111
Toole, Lucy Hicks. See DeLaine, Lucy Toole

Trading. See Bartering
Transportation, 86, 96
Tucker, Samuel, 31
Turner, Mary Wiggins, 76
Tuskegee Institute, 16, 141, 143, 161

U.S. Department of Agriculture (USDA), 15, 109, 146–47, 161

Vagrancy laws, 146
Vegetables, 61. See also Gardening

Wade, Lucy, 148
Ward, John, 57–59
Washington, Booker T., 154
Watkins (J. R.) Medical Company, 29, 41–44
Wayne County, N.C., 139–40, 169
Welfare, 156–57, 169. See also Public health
Whites: racial attitudes of, in North Carolina, 149–51
White supremacy, 29, 32, 46, 47, 149–50
Wilkes County, N.C., 102–3
Williams, Sarah J., 66, 148, 151, 155, 156–57
Wilson, Woodrow, 142
Winski, Arnalee, 8
Women. See Commodities, women's production of; Farm women; Home demonstration agents; Home demonstration clubs
"Work or fight" laws, 146
World War I, 145–47
World War II, 104, 175

Yeomanry, 14, 187 (n. 21)
Young, Hattie, 74–75
Young, Walter, 74–75